Economics of Water Resources
The Contributions of Dan Yaron

NATURAL RESOURCE MANAGEMENT AND POLICY

Editors:

Ariel Dinar
Rural Development Department
The World Bank
1818 H Street, NW
Washington, DC 20433

David Zilberman
Dept. of Agricultural and
Resource Economics
Univ. of California, Berkeley
Berkeley, CA 94720

EDITORIAL STATEMENT

There is a growing awareness to the role that natural resources such as water, land, forests and environmental amenities play in our lives. There are many competing uses for natural resources, and society is challenged to manage them for improving social well being. Furthermore, there may be dire consequences to natural resources mismanagement. Renewable resources such as water, land and the environment are linked, and decisions made with regard to one may affect the others. Policy and management of natural resources now require interdisciplinary approach including natural and social sciences to correctly address our society preferences.

This series provides a collection of works containing most recent findings on economics, management and policy of renewable biological resources such as water, land, crop protection, sustainable agriculture, technology, and environmental health. It incorporates modern thinking and techniques of economics and management. Books in this series will incorporate knowledge and models of natural phenomena with economics and managerial decision frameworks to assess alternative options for managing natural resources and environment.

Water is a scarce resource. Therefore, the efficient use of water is an economic issue. When scarcity is characterized by both quantity and quality, the economics becomes even more complicated. This book addresses the economic aspects of water quantity and quality interaction. The book offers a series of analytical and policy chapters that cover field, farm, and regional-level water allocation issues, as well as international aspects of water sharing. The book demonstrates the ability and constraints of economic approaches in solving various water allocation problems.

The Series Editors

Recently Published Books in the Series

Feitelson, Eran and Haddad, Marwan
Management of Shared Groundwater Resources: the Israeli-Palestinian Case with an International Perspective
Wolf, Steven and Zilberman, David
Knowledge Generation and Technical Change: Institutional Innovation in Agriculture
Moss, Charles B., Rausser, Gordon C., Schmitz, Andrew, Taylor, Timothy G., and Zilberman, David
Agricultural Globalization, Trade, and the Environment
Haddadin, Munther J.
Diplomacy on the Jordan: International Conflict and Negotiated Resolution
Just, Richard E. and Pope, Rulon D.
A Comprehensive Assessment of the Role of Risk in U.S. Agriculture

Economics of Water Resources
The Contributions of Dan Yaron

Collected by

Ariel Dinar

The World Bank, Washington DC, U.S.A.

And

David Zilberman

University of California, Berkeley, USA

SPRINGER SCIENCE+BUSINESS MEDIA, LLC

Library of Congress Cataloging-in-Publication Data

Dinar, Ariel.

Economics of water resources : the contributions of Dan Yaron / collected by Ariel Dinar and David Zilberman.

p. cm. – (Natural resource management and policy ; 24)

Includes bibliographical references and index.

ISBN 978-1-4613-5294-5 ISBN 978-1-4615-0899-1 (eBook)

DOI 10.1007/978-1-4615-0899-1

1. Water resources development—Economic aspects. 2. Water-supply--Economic aspects. 3. Water quality management—Economic aspects. I. Dinar, Ariel, 1947 – II. Zilberman, David III. Title. IV. Series.

HD1691 .Y37 2002

333.91—dc21 **2002019827**

Printed on acid-free paper.

Dan Yaron

July 17, 1922 - December 26, 1999

CONTENTS

LIST OF FIGURES

LIST OF TABLES

ACKNOWLEDGEMENTS

Many friends and colleagues provided essential input and helped to bring this book to fruition. Nitza Sadeh re-typed several of the book chapters and helped update Dan's list of graduate students; along with Aaron Ratner, Eli Feinerman, Yoav Kislev, Yoram Porat, and Yakir Plessner.

The publication of this book was co-sponsored by a grant from The Center for Agricultural Economic Research, Rehovot, Israel.

Preface

Colleagues and friends of Dan Yaron submitted the following tributes. While each submission comes from an individual who knew Dan in a very different way, they all remark on his immeasurable contributions to the field of agricultural economics, his unique approach, which combines his training and experience with his scientific background, and the admirable professionalism and civility that was apparent in every project he undertook. His work, initially inspired by the challenge of farmers in the arid Negev, eventually took him to the United States to work with universities and to serve on commissions furnishing his research with global applicability. Dan is not only admired for his enormous contributions to the vast body of research available in his field, but also for the commitment and dedication he epitomized. He will be greatly missed by those of us who were fortunate enough to make his acquaintance.

ELI FEINERMAN
The Hebrew University of Jerusalem, Rehovot, Israel

Dan Yaron, my teacher and mentor of blessed memory, was a man of wisdom, thought, counsel and deeds. His many talents, his endless energy and his ambition led him to blaze new trails in research, to ask the relevant questions while separating the wheat from the chaff, and to answer them while mastering the most advanced scientific analyses.

In his scientific work Dan was a man of vision, patient enough to go into the smallest details, and capable of shaping new ideas from meticulous study of the facts. He was thorough and persistent, eager to learn, and driven by a tireless need to help farmers solve their economic difficulties. He worked hard to build bridges between academic theories and their potential application in the field. He was prepared to fight for his professional beliefs and his social ideas and did not hesitate to express his views in a loud, clear voice, even when they were unpopular, backing up his opinions by facts and data.

Dan was also an outstanding teacher. Many of his former students are at the forefront of the Israeli economic scene, in both the public and the private sectors. Dan's specific contribution to their fund of knowledge, work habits and drive was immense. Many of us had great admiration for him. We accepted his leadership and often consulted with him, trusting his knowledge and wisdom, and drawing strength and encouragement from him.

Dan was a modest man in his dress and lifestyle. As a result, and also because of his direct approachability, everyone called him simply "Dan"—it was seldom that his title of "Professor Dan Yaron" was heard. He was said to be strict, and indeed he demanded much of those who worked with him, but not more than he demanded of himself. I also knew Dan as a good hearted and emotional person, who endowed with a sense of humor, a man who enjoyed jokes and everyday conversation. He had the gift of listening, and would help and assist without expecting any reward. As a father he was very proud of his sons, and as a husband he was totally devoted to his wife Giza.

Dan Yaron was a complex man with great influence on the scientific community and the society in which he lived. He embodied different, seemingly contradictory traits. He could grasp the broad picture without missing any of the details; he was extremely diligent and hardworking, but loved company and conversation; he demanded much of his subordinates, but respected their freedom of choice; he adhered strictly to economic theories in his work, but was also pragmatic and practical; he was an idealist but did not lose himself in imaginary spheres.

Dan's passing was a loss to his family, his pupils, his associates, the Israeli farmers and the scientific community. May his memory remain.

YOSI SHALHEVET
Agricultural Research Organization, Bet Dagan, Israel

Dan Yaron had been a friend for many years. Our friendship went beyond our professional association, though it began through it. It was a pleasure to be with him and his wife, Giza, to talk politics and discuss personal issues or, to have a cup of coffee and, of course discuss business. Our professional and academic interests intersected on the problems of efficient water use, but especially on the use of brackish water for irrigation.

Dan was an economist who initially knew little about soils and crops in relation to water and salinity. That is where I came in, together with our good friend, the late Eshel Bresler. It was not long before Dan became knowledgeable in our field of expertise. That not withstanding, we continued our professional association and Dan never took the liberty to pursue issues of mutual concern without consulting us, though the consultation was at times redundant as he could have done it by himself. When it came to professional integrity and competence Dan did not compromise or take shortcuts. His work on the economics of water use in agriculture, especially the interaction with salinity, made him a foremost authority on the subject and earned him a prestigious position on national decision making and consulting committees.

It was during my involvement with the issue of the decision by the Hebrew University administration in Jerusalem to close down the Department of Agricultural Economics in Rehovot, that I came to appreciate Dan's character, determination and perseverance. The administration of the Hebrew University decided that the learning of agricultural economics as an independent field of study was unnecessary. Agricultural economics is economics, they said, and should be studied in the Department of Economics in Jerusalem. Dan Yaron did not agree with this point of view and decided to fight it.

To aid him in the struggle he enlisted a group of people who, he thought, could help convince the Hebrew University's authorities that the Faculty of Agriculture in Rehovot needed a Department of Economics. Furthermore, he felt that students of agricultural economics should study in an ambience permeated with agriculture. At the time, I was the director of the Agricultural Research Organization and the Chief Scientist of the Ministry of Agriculture, and I agreed with him. Together with other people in key positions we formed a committee, inspired by Dan's relentless leadership, and managed to reverse the decision. As the saying goes "perseverance pays" and it was Dan who persevered. The Department of Agricultural Economics owes its independent existence in Rehovot to Dan Yaron.

Let his memory be with us forever.

URI SHAMIR
Water Research Institute, Technion-Israel Institute of Technology, Haifa, Israel

We had a long relationship, always warm and mutually respectful. I worked with Dan for several years at Mekorot Water Company, on the development of a model for annual operation of Israel's National Water System. This required a comprehensive view of the entire system, and Dan had a knack for determining what is essential and important, and what is feasible. We then shared ideas and initiatives and called on each other to join in on the other's activities.

As Director of the Water Research Institute (WRI), I have the privilege of suggesting to the President of the Technion the names of candidates for the WRI's International Science Advisory Committee, which is constituted of leading international experts in various areas of water science, technology, engineering and management. For several years I tried to get Dan to agree to be a member, knowing full well that he would take the charge seriously and execute it with dedication and wisdom. But at that time his family required Dan's full attention, and he did not wish to undertake a responsibility that would require him to be absent.

When I again approached him to ask whether he would reconsider, he thought it over, but finally declined, with a typical Dan-apology for supposedly letting a friend down.

We all miss Dan.

CHARLES (CHUCK) W. HOWE
Institute of Behavioral Sciences, University of Colorado at Boulder, Boulder, CO, USA

I had the privilege of knowing and working with Dan Yaron on several occasions. Along with all of our profession and beyond, I have benefited from his extensive applied and theoretical work and have taken inspiration from his devotion to scholarship and practical problem solving. In his presence, one felt his strength of character, his intellect and his personal warmth. Fortunately for us in the United

States, Dan frequently spent leave periods here at top universities including North Carolina State, Harvard, Iowa State, Colorado State and the University of Chicago.

Dan was one of the world's leading experts on the effects and control of salinity in irrigation water and soils. His publications in this area, both books and papers, comprise a large part of the scientific literature. Much of this work was, quite naturally, aimed at problems in Israel, but his work has much wider applicability in the arid and semi-arid regions of the world. I had the pleasure of arranging his participation in an early workshop, *Salinity in Water Resources,* held in Boulder in 1973 under the sponsorship of the U.S. Office of Water Resources Research and other agencies. Dan's edited book, *Salinity in Irrigation and Water Resources* (1981), has been a standard reference since its publication.

Dan made significant contributions to several fields, nearly always with a clear application to the problems of his country: strategies for rural development and optimal selection of rural industrial activities; technological diffusion and the modernization of traditional agriculture; and, needless to say, the optimal allocation of water and water quality management in arid areas.

We are all indebted to Dan Yaron for his scholarship and his dedication: to problem solving; to his country; and to his family.

ROBERT A. YOUNG
Colorado State University, Ft. Collins, Colorado, USA

I am pleased to add my reflections to this remembrance of Dan Yaron. His creative work on science and policy related to water first came to my attention early in my career. Subsequently, our common interests led to (too brief) formal collaborations on research, first when he came on leave to Colorado State University and later in Israel.

Dan worked at the forefront of a generation of economists who introduced the application of neoclassical economic theory, together with new and evolving quantitative and computational methods, to study problems of agricultural production in arid climates; particularly those relating to the use of irrigation water in agricultural crop production. (Prior to the 1960s, most agricultural economists who studied agricultural water management and policy issues were trained in the then-influential Institutionalist School of Economics, which combined a mistrust of abstract theory and quantitative analysis with a primary interest in the role of institutions in the economy).

To the emerging area of economic research on economic relationships in irrigated crop production, Dan brought a unique combination of training and experience. His early preparation in production agriculture provided him with a strong background in biological and physical sciences applied to agriculture, and facilitated his subsequent innovative interdisciplinary work with applied agricultural scientists. Moreover, Dan's aptitude for applied mathematics enabled him to readily translate hypothesized relationships into formal mathematical models. Thus, he led the way to formulating and implementing models of crop response to irrigation water quantity and timing. His path-breaking body of work on the effects of salinity on crop productivity--a universal problem in arid lands agriculture--was of singular importance.

Dan also emphasized the neoclassical economic approach to analyzing water policy. Although acknowledging unique interdependencies and externalities, this methodology contended that water--along with land, labor and capital--is best treated as a normal productive input, subject to the laws of diminishing marginal returns and characterized by a demand which is responsive to cost or price. Early on, he challenged policies, which encouraged low-valued water uses and pointed out the adverse long-term implications of public programs that under priced the water resource.

But, research was not Dan's only focus. Over the years, he devoted uncounted hours to public service, particularly serving on numerous advisory committees. I once sat through a graduate seminar he conducted, and learned first-hand why his students were so appreciative. He and his wife were gracious hosts to many who, like myself, came to visit and work with him in Israel. His was certainly a career to be admired and emulated.

VERNON W. RUTTAN
University of Minnesota, St. Paul, Minnesota, USA

I have known and have drawn on the research of Dan Yaron during most of my professional career. My initial interest in Dan's research stemmed from my own research on the economic demand for irrigated acreage in the United States in the early 1960's. I have continued to follow his research, particularly on the economic and institutional aspects of water resource allocation and use.

I did not meet Dan Yaron, however, until 1990 when he spent part of a sabbatical at the University of Minnesota. We had met informally a number of times to discuss his research on indigenous Arab family farms in the Nazareth Region of the North of Israel. We also discussed his plans for closer collaboration between Israel and Egypt in the area of water resource allocation and use.

I was particularly impressed with the effort that Dan had made to work with Arab agricultural extension workers on problems of agricultural resource use, production and marketing. Dan was particularly interested in the adoption of innovations and the responsiveness of farmers to economic opportunity.

Among the important research findings of Dan and his colleagues in the Nazareth area, was that farmers with only elementary education were capable of adopting and managing complex technologies if proper extension support was made available.

In addition to our shared interests in technical and institutional change, I found Dan to be a wonderful friend and colleague during his stay at Minnesota.

RONALD G. CUMMINGS
Georgia State University, Atlanta, Georgia, USA

I have met Dan Yaron on several occasions, and found him to be a warm and open man. The great respect that I have for Dan derives from my reading of his published research, and from the influence of that research on my own professional development. I will then briefly comment on his works that, in my view, represented particularly important contributions to the state of the art in water resources management.

I have always viewed Dan's published research as pushing the intellectual envelope along two related directions: the advancement of methodology; and pioneering works related to water quality and the agricultural sector. In terms of methodological contributions, one would certainly not be surprised to see one of Earl Heady's students using linear programming (LP) as a tool for exploring crop response functions. But Dan was one of the early pioneers in the search for means by which the limitations of LP might be overcome by the conjunctive use of LP with other computational algorithms or methods. Thus, in the early 1970s one finds a series of papers, which explore the efficacy of various alternative computational methods for policy analyses. Papers that I found to be particular helpful for my early efforts to model water resources systems include: Dan's combination of LP and sequential programming techniques to examine farm growth and capital accumulation under conditions of uncertainty; the conjunctive use of LP and decision tree analysis for assessing long-run farm planning; and his use of an integrated system approach applied to the problem of determining irrigation policy under conditions where rainfall is unstable. In my efforts to follow Oscar Burt's lead in the use of dynamic programming techniques to issues related to the scheduling of irrigation, I found Dan's 1973 paper with A. Olian to be particularly helpful ("Application of Dynamic Programming in Markov Chains to the Evaluation of Water Quality in Irrigation"). Dan's work during this period accomplished a great deal in terms of demonstrating how emerging analytical systems could expand the breadth and depth of policy analyses of contemporary importance to the agricultural sector.

In terms of policy analyses, Dan was one of the early scholars to become interested in the problems of water quality as they arise in the agricultural sector. His 1972 paper with Bresler added new dimensions to the fledgling environmental economist's concerns with water quality. This (in my view) seminal work introduced the importance of leaching fractions vis-à-vis the accumulation of salts in soils and their implications for optimal soil water regimes. This work is then followed by a series of papers that expand the scope of analyses to include such things as seasonal aspects of water quality control , irrigation scheduling, spatially variable fields, the possibility of mixing irrigation water, and the value of information on a crop's response to soil salinity. In closing, I invite the reader's attention to Dan's more recent work with Ariel Dinar that explores innovative approaches to the use of effluent for irrigation.

I know that Dan's friends will deeply miss him. Many like me owe Dan an enormous debt of gratitude for the influence of his work on our lives. I will certainly miss his presence in our profession.

1

DAN YARON: THE PERSON, HIS WORK AND HIS LEGACY

Ariel Dinar
World Bank, Washington, DC, USA

David Zilberman
University of California, Berkeley, California, USA

Dan Yaron was born in Poland and immigrated to Israel with his family in 1935. His teenage years were spent in Tel-Aviv. He was a member of Kibbutz Manara between 1946 and 1951, prior to returning to school to earn a Bachelor of Science and Master's degrees in Agriculture at the Rehovot campus of the Faculty of Agriculture of the Hebrew University of Jerusalem. Yaron completed his Ph.D. in Agricultural Economics at Iowa State University, under the supervision of Earl Heady, in 1960. Upon his completion, Yaron returned to the newly established Department of Agricultural Economics and Management of the Hebrew University of Jerusalem in Rehovot.

Dan Yaron's broad contributions to the fields of farm management, production economics, and water resource economics have affected the focus and direction of the work of many of his colleagues around the world.

"Dan can be viewed as the founder of modern water resource economics, and was ahead of his peers by 25-30 years."

"A number of his articles have become classics in their field."

"The hallmarks of Dan's published work [were his]... imagination and innovation. These characteristics are clearly seen in Dan's work, which introduced the use of economic analysis to water resource management problems."[1]

Yaron's great strength was his ability to combine policy-oriented research with practical experiences. Starting as a farmer on a kibbutz in northern Israel, and continuing as a student of agriculture at the Hebrew University, he realized the importance of direct links between real-world problems and academia. This realization guided Yaron throughout his career as a teacher and researcher at the Department of Agricultural Economics and Management at the Hebrew University of Jerusalem, and in his work with public agencies.

"Professor Yaron is widely known, not only for his technical economic research, but for his concern with application." The best word to describe Yaron and his work is 'relevance.' The drive for relevance in his analytical and policy-oriented research is reflected in his list of publications, and distinguishes all that he has done.

Yaron was a pioneer and he understood the critical link between the natural sciences and agricultural economics. He initiated many collaborative research projects with soil scientists, irrigation specialists, agronomists, engineers, and experts from other disciplines to produce the information relevant for decision-making models. His works with Bresler, Bielorai, and Shalhevet are widely cited. Moreover, recognizing relative advantages and deficiencies in the level of knowledge in various fields of agricultural economics in Israel, Yaron initiated many joint projects with American economists in order to exchange knowledge and experience.

While Yaron's research and teaching career encompassed several fields, his analytical and empirical contributions in water resource economics and farm management are the most noteworthy. Starting in the early 1960s, he developed and applied concepts related to water-yield response functions.

"He did pioneering work on the introduction of water quality into crop production functions, as well as formulating dynamic economic models for optimal inter-temporal allocation of irrigation water, including quality issues." These economic tools were incorporated into farm and inter-regional economic models that have been used for policy and management purposes. Yaron was also a pioneer, introducing the concept of the value of marginal productivity of water and incorporating it into the decision-making processes of both farm managers and agricultural water planning institutions in Israel. He affected agricultural policy making, in both agricultural farm planning and water resources development, by providing a rich set of empirical evidence to back up his analysis.

Dan Yaron was involved in shaping the research agenda in water resource economics. He served as a member of the editorial councils of *The Israel Economic Quarterly*, *Water Resources Management*, and *Oxford Agrarian Studies*. He chaired numerous national and international committees for funding research projects. Yaron also influenced the decision-making process of the water sector in Israel throughout his advisory positions and his membership roles with Mekorot-Israel's Water Company, Tahal-Israel Water Planning Company, Israel Water Commission (on various committees for planning, policy, water quality, and low water quality).

Yaron's research brought him to North Carolina State University, Harvard University, Iowa State University, Colorado State University, the University of Chicago, Oxford University, and the University of Minnesota where he worked with leading contemporaries like Dorfman, Heady, Howe, Young and Tolley. Yaron's work with these colleagues during sabbatical periods extended to further collaboration and affected the direction and extent of the research produced during this time. This book attempts to draw on the extensiveness of Yaron's most collaborative work.

In the 1960s and 1970s Yaron developed, introduced and applied mathematical programming approaches to planning purposes at farm, regional, and national levels. His contributions included the application of mathematical optimization techniques such as Separable planning, quadratic planning under competitive equilibrium, dual-primal analysis, and incorporation of income effects into planning models (e.g., Yaron and Heady, 1961; Yaron et al., 1965; Yaron, 1967; and Yaron and Horowitz, 1972a, b).

"His applications of mathematical programming to agricultural management, in general, and to irrigation water allocation, in particular, were at the cutting edge of work in those areas."

In the late 1970s and early 1980s Yaron attained worldwide recognition for his scientific contribution to the development and application of mathematical models to the optimal management of water quantity-quality problems. Additional recognition for his contributions are reflected in the wide citation of several of his publications. Examples include: Yaron and Bresler (1970) and Yaron and Olian (1973), with more than 500 citations, and Yaron et al. (1980), with more than 300 citations.

"His contributions to the (still) very difficult problem of optimizing interseasonal water applications stand out."

In the last decade, realizing the important role of marginal water sources in alleviating water scarcity, he focused his efforts on several issues that, at that time were, by themselves, pioneering. They include the economic evaluation of the substitution of water with varying qualities (e.g., Feinerman and Yaron, 1983), and the regional cooperation in sharing and managing marginal water (saline water and municipal waste water). In the course of regional cooperation, Yaron and his students modified and applied cooperative and noncooperative game theory approaches to water resources issues. These works yielded several important contributions to the literature that present strong empirical evidence of the important role of cooperation among interest groups in solving water quality and scarcity problems.

Like many others, but perhaps more than anyone else, Yaron realized the terrible cost of war and appreciated the need for peace. He believed in the role that cooperation played among the nations of the Middle East. Concurrent with the peace negotiations between Egypt and Israel, Yaron initiated and led a long-term USAID-funded project aimed at sharing and transferring agricultural technologies between Egypt and Israel in order to promote agricultural progress. The underlying philosophy of the project, similar to Yaron's professional belief, was that strong links among applied scientific research, economic analysis, and on-farm implementation (involving the relevant supporting disciplines) are required for effective adoption of new technologies, and consequently for improving the well-being of the people in the region. The project demonstrated the usefulness and importance of agricultural economic principles in regional development.

The project yielded several significant contributions to the literature, which highlight the effects of input quality on the adoption of modern irrigation technologies. In addition to its excellent academic and policy oriented results (reflected in many publications and results on the ground) and joint Israeli-Egyptian and Israeli-Palestinian publications (e.g., Regev et al., 1990; El-Rayes et al., 1995), the seven-year project also created strong personal relationships among researchers in Egypt and Israel, and personal relationships between Palestinian and Israeli researchers who joined the project. The professional and personal relationships have continued, and have been "essential in laying the foundation for future development in the region."

In addition to these research accomplishments, Yaron distinguished himself as an excellent teacher. He supervised nearly fifty graduate students (see Appendix), an impressive number by Israeli or international standards. Many of his students carried his ideas as they served in key decision-making positions in government and private sector organizations and continue to do so.

Yaron served as Chairman and Head of the Department of Agricultural Economics and Management at the Hebrew University of Jerusalem for several periods. During his tenure, in 1988-1990, he successfully led a campaign against an administrative decision to abolish that department. His enormous efforts eventually demonstrated the vital role of the Department of Agricultural Economics and generated a surge of recognition of the importance of the profession.

Dan Yaron has distinguished himself as a researcher, a teacher, an administrator, and public servant and has established an outstanding record in each field. "He was the intellectual giant in the economics of water quality for irrigated agriculture. His academic leadership, his contributions to advancements in water and agricultural economics, and guidance in water policy have made him one of the world's best known and most authoritative water resources and agricultural economists."

"Our profession is indebted to Yaron for his intellectual leadership in the area of water resource management in irrigation." We owe him so much, not only for his continuous work on mainstreaming water resource economics, but for the personal qualities he was able to bring to his research and educational activities.

This book represents the array of issues that were addressed in Yaron's work in water economics over the past four decades. The various chapters do not only reflect the direction and the methodological development of his research, but also provide a glimpse into the kind of person Dan Yaron was and the strong bonds he had with his peers and students alike.

The second chapter highlights the important policy issues associated with the Israeli water sector. It introduces the reader to Yaron's distinct philosophy on water policy, which involved a comprehensive approach, including economic incentives, technology, and the use of various water sources. Chapter 3 represents a significant portion of the early days of Yaron's work. It demonstrates the application of production functions to the estimated demands for irrigation water at farm, regional and national levels. Production functions were also used by Yaron to derive policy recommendations regarding the timing, quantity and quality considerations involved in irrigation. Chapter 4, coauthored with two graduate students, Stratiner and Weisbrod, and an irrigation scientist, Shimshi, addresses the problem of unstable rainfall in an arid region. Chapter 5, coauthored with soil scientist Bresler focuses on the economics of water quality and salinity. It also lays out the building blocks for following collaborative work with soil scientists and economists.

Several other chapters in the book are also based on the concepts in Yaron-Bresler models. The next three chapters apply dynamic programming methods to problems of water quantity and quality in irrigation. Chapter 6, coauthored with Olian, also a former student of Yaron, is one of the first attempts to incorporate water quality considerations within a dynamic economic framework. Chapter 7 develops and applies a dynamic programming model for water quality application on a daily basis and is based on Harpenist's master's thesis (written in collaboration with Bresler and Bielorai). In Chapter 8, a system based on a linear programming model and a dynamic programming model is developed to optimize water allocation on a farm during peak seasons of cotton field irrigation. The approach, developed with Dinar, at that time an Agricultural Extension Agent, was later extended to an extension project in the Rehovot region in Central Israel.

The value of information is explored in Chapter 9, coauthored with Feinerman, Yaron's first Ph.D. student. Again, here, the concept of a response function to water quality is used to derive the value to farmers and decision makers of investing in acquiring additional information on the relationship between water quality and crop yield.

In Chapter 10 a model of the seasonal aspects of water quality in a river basin is presented. This chapter illustrates the foundation that Yaron established for further work on the economics of regional wastewater reuse that did not gain momentum until several years later. The concept presented in this chapter was initiated while Yaron spent his sabbatical at the University of Chicago, collaborating with Tolley. Chapter 11, coauthored with Dinar,

Yaron's second Ph.D. student, deals with economic optimization of municipal wastewater treatment and reuse in irrigated agriculture. The model is built on regional cooperation principles, and highlights not only the economics of size, location, and treatment level, but also the nature of cost allocation arrangements between the wastewater producers and users. In Chapter 12, coauthored with Ratner and Dinar, the authors generalize, based on Ratner's masters thesis and Dinar's Ph.D. thesis, regarding the suitability of cooperative game theory in addressing water resources problems.

Chapters 13 and 14 mark a new focus in Yaron's work, reflecting his concern for regional cooperation between Israel and its neighbors. Chapter 13, coauthored with Dinar, is a representation of the long-term USAID project (mentioned above) on transfer of technologies between Egypt and Israel. It addresses the adoption and abandonment of irrigation technologies in various regions in Israel and Gaza. In Chapter 14, Yaron assesses the water situation in reference to the Palestinian entity and Israel, and reaches the unavoidable conclusion that the only sustainable solution to a long-run water allocation arrangement is for the two to cooperate. Yaron also suggests the framework for non-cooperative agreements and shows the welfare loss compared with suggested cooperative arrangements.

The conclusion chapter of the book attempts to place Yaron's work within the body of published scientific literature. It is seen very clearly where and how was the research agenda in the world affected by the methodologies developed and used, and by the findings in Yaron's work.

This book aims to highlight the vast scope of Yaron's work in water economics, which combines several conceptual layers that both lead to and build on each other. It consists of field-level water allocation models with water quantity and quality considerations and farm-level and regional-level models with similar considerations. In its entirety, Yaron's work incorporates both technical and policy considerations and is most relevant to decision makers at all levels.

REFERENCES

Feinerman E. and D. Yaron. The value of information on the response functions of crops to soil salinity. *Journal of Environmental Economics and Management,* 10:72-85. 1983.

Mohammed, H. El-Rayes, Salah M. Faroukh, Dan Yaron and Hillary Voet. *Development of Agriculture in the Gaza Region.* Center for Agricultural Economics Research. The Hebrew University of Jerusalem. Rehovot, Israel. March, 1995.

Regev, A., A. Jaber, R. Spector and D. Yaron. Economic evaluation of the transition from a traditional to a modernized irrigation project. *Agricultural Water Management,* 18:347-363. 1990.

Yaron, D. and E. O. Heady. Approximate and exact solution to non-linear programming problem with separable objective function. *Journal of Farm Economics,* 43:57-70. 1961.

Yaron, D., Y. Plessner and E. O. Heady. Competitive equilibrium - application of mathematical programming. *Canadian Journal of Agricultural Economics,* 13:65-79. 1965.

Yaron, D. Incorporation of income effects into mathematical programming models. *Metroeconomica,* 19:141-164. 1967.

Yaron, D. and U. Horowitz. A sequential programming model for growth and capital accumulation of a farm under uncertainty. *American Journal of Agricultural Economics,* 54:441-451. 1972a.

Yaron, D. and U. Horowitz. Short and long run farm planning under uncertainty - integration of linear programming and decision tree analysis. *Canadian Journal of Agricultural Economics,* 20:17-30. 1972b.

Yaron, D. and A. Olian. Application of dynamic programming in Markov chains to the evaluation of water quality in irrigation. *American Journal of Agricultural Economics,* 55(3):467-471. 1973.

Yaron, D. E. Bresler, H. Bielorai and B. Harpenist. A model for optimal irrigation scheduling with saline water. *Water Resources Research,* 16:257-262. 1980.

---

[1] Citations taken from letters of colleagues.

2

THE ISRAEL WATER ECONOMY: AN OVERVIEW*

Dan Yaron
The Hebrew University of Jerusalem, Rehovot, Israel

1. WATER POTENTIAL

There are several estimates of Israel's water potential; they range from 1,517 to 1,781 million cubic meters (m^3) per year, not including Gaza and the South Jordan Valley aquifers. The detailed estimates are presented in Table 2.1. The potential of water from natural sources includes about 160 million m^3 of brackish water, defined as water including more than 400 parts per million (ppm) chlorides (Cl), roughly equivalent to 1,000 total dissolved solids.

It should be noted that: a) the water potential depends on land use—the type of vegetation and the extent of urban uses; b) there is a difference between the hydrological potential (which includes all the sources of water) and the practical potential (which represents the quantity of water practically available for use under the prevailing economic conditions); c) the figures mentioned above are expected values. The coefficients of variation of the natural replenishment of the major watersheds in Israel are Kinneret Basin, 0.36; Coastal Plain, 0.29; Yarkon Taninim (part of the Mountain Aquifer), 0.22; all three watersheds combined, 0.27 (Schwartz, 1990).

The policy options faced by the policymakers are either to maintain a constant water supply from natural sources over the years with a low expected value and a low standard deviation or to maintain a flexible water supply with a relatively higher water potential and higher standard deviation, as well. This issue is closely related to the structure of agriculture and its crop mix.

---

*Permission to publish this chapter was granted by Kluwer Academic Publishers. The chapter was originally published with a similar title in *Decentralization and Coordination of Water Resources Management*, (D. Parker and Y. Tsur, eds.). 9-22, 1997.

Table 2.1. Water potential from natural sources, excluding Gaza and South Jordan Valley (million m^3 per year)

Version	(1)	(2)	(3)
Source:			
Kinneret Basin	660	660	660
Ground and flow Water	1,195	980	1,063
Total	1,855	1,580	1,723
Conveyance losses	75	63	69
Net potential	1,781	1,517	1,654

Sources: (I) Tahal Consulting Company (1988), (2) Nevo (1992), and (3) based on Hydrological Service

2. THE BALANCE OF WATER

The projected water use in Israel in the twenty-first century depends on the projection of population and water use per capita. The recent projection regarding population for the year 2010 is 6.9 million inhabitants in Israel including the settlements in the West Bank and South Gaza Strip. The quantity of water projected for domestic use is approximately 700 million m^3 per year with an additional 140 million m^3 for industrial use. The projection for domestic use is based on the assumption of 100 m^3 per capita in 2010 and the years beyond. The current average use per capita is about the same.

The reason for using 100 m^3 per capita and not a higher figure, which could reflect higher standards of living, implicitly assumes water policy aimed at lower use levels. In effect, in years of short supply, the use per capita can be even lower than 100 m^3 of proper administrative restrictions take place.

Table 2.2 shows the totals of the projected urban use of water, which, when compared with the potential of water from natural sources, give the residual available to agriculture. Note that this residual includes about 160 million m^3 of brackish water. The last row of Table 2.2 presents the potential for reclaimed wastewater, which jointly with the residual in row 5 could be allocated to agriculture. The figures for the years 2030 and 2040 are extrapolations that illustrate the potential situation during the first part of the twenty-first century.

Table 2.2 suggests that an increased share of water supply to Israeli agriculture would be based on low-quality water (brackish and reclaimed wastewater). According to this projection, already in the third and fourth decades of the twenty-first century, the shortage of fresh water to Israeli agriculture will be quite severe. Note that deviations from the projections in Table 2.2 are very likely, due to variation in the potential, and different rate of growth of population and use per capita. The table presents a general simplified view. A real-life planning should follow a probabilistic approach.

Table 2.2. Urban water use and residual for agriculture from natural sources (million m^3 per year)

Year	1990	2010	2030	2040
Domestic[a]	482	700	980	1,160
Industry	105	140	150	150
Total (A)	588	840	1,130	1,310
Net Potential[b] (B)	1,654	1,654	1,654	1,654
Residual[b] (B)-(A)	1,066	814	524	344
Wastewater Potential	240	400	540	630

[a] 100 m^3 per capita beyond 1990
[b] Including 160 million m^3 brackish water

It should be emphasized that the figures in Table 2.2 avoid, on purpose, the discussion of the claims for water by the Palestinian Authority, due to its sensitivity. Here we shall be satisfied with three comments: (1) the dispute over water between Israel, the Palestinian Authority, and the Hashemite Kingdom of Jordan should be resolved around a negotiation table; (2) data like those presented in Table 2.2 may serve as an input to the negotiations (see also Yaron, 1994), and (3) any quantity of water transferred from the current or projected use in Israel to its neighbors will primarily affect the agricultural sector of Israel.

A schematical presentation of the map of Israel and the core elements of its water system are shown in Figure 2.1. The major aquifers and the National Water Carrier (NWC), which conveys water from Lake Kinneret to the South and the Negev, are presented as well as the borders with the neighbors. The borders with the Palestinian Authority and Syria are under negotiations at the time of this writing (These statements were valid for 1997).

3. WATER SALINITY PROBLEM

There are two aspects of the salinity problem. The first one refers to local spots of brackish water in certain regions, which together amount to about 160 million m^3 per year.

The second, and considerably more important problem in the long run, is the strong trend faced by Israel of increasing salinity over time in most of its natural water sources. This process is the result of (1) reduction of natural drainage and natural salt leaching to the sea, due to the very intensive exploitation of Israel's water sources; (2) intrusion of sea water in some locations along the coastal plain; (3) import of salts with irrigation water from Lake Kinneret to the regions served by the NWC. (Even though the salt content of Lake Kinneret, the source of NWC is relatively low (200 to 240 ppm Cl), the salt brought in by the NWC gradually accumulates in the soil and ultimately percolates to the groundwater); (4) irrigation with wastewater, which is more saline than fresh water.

At the farm level the focus is on the optimal use of brackish water in the short and long run (Yaron, 1984; Feinerman and Dinar, 1991).

At the national/regional level the major problems are (1) how to deal with the externalities of irrigation with brackish water and (2) how to incorporate brackish water into the farmers' water quotas (as long as this system prevails) and what should be the rate of substitution between the brackish and the high quality water. A Water Commission Committee has recently issued policy recommendations in this regard, which are being put into effect. Details fall beyond the scope of this chapter (These statements were valid for 1997).

4. TREATED WASTEWATER AND ITS USE

As shown above, treated wastewater will continuously comprise an increasing share of the total supply potential to agriculture and perhaps to industry as well. It is assumed that between 50 to 60 percent of household water can be recycled and reused, if there is sufficient demand for its use.

The alternatives regarding wastewater are: (1) disposal to the sea, (2) treatment and reuse by agriculture, and, perhaps (3) reuse by industry. Note that reuse by households and offices is not currently considered as a viable alternative even though it may become such in the future (with dual supply systems in households and offices, as, for example, in large office buildings in Japan). According to the international agreements, disposal of wastewater to the sea implies a certain level of treatment (the base level) and long pipes for offshore discharge.

The use of treated wastewater in agriculture involves strict environmental rules, with a double aim: preservation of groundwater when irrigation takes place above unconfined aquifers, and public health considerations aimed at preventing the spread of bacteria, viruses, carcinogenic materials, and so on. Strict restrictions on the use of treated wastewater in irrigation have been passed; the issue of gradual contamination of groundwater by chemicals is being studied (see, for example, Falkovitz and Feinerman, 1994; Feinerman and Voet, 1995); however, problems of enforcement of the regulations are far from being solved (see Schwartz, 1990; Avnimelech, 1991).

A major issue is agricultural use versus disposal. According to a recent report to the Water Commission (Shevah and Shelef, 1993): "If the environment preserving regulations are followed, allocation to agriculture seems economically viable in most cases." A major difficulty involved in the allocation of treated wastewater to agriculture is the problem of interseasonal storage. Underground storage capacity is limited, and even if locations can be found, there is an alternative cost of storing treated wastewater rather than fresh water; storage in open ponds bears too an alternative cost in terms of land. Furthermore, there is a loss in quantity due to evaporation, and even more important is the effect of evaporation on increasing salinity. Available observations suggest that the salt content of reclaimed sewage is higher by about 100 ppm Cl than the incoming fresh water supplied to urban consumers. If the current salt content of the water supply by the National Water Carrier is about 220 ppm Cl, the salt content of reclaimed sewage without underground mixing will be 320 ppm and after open-pond storage may reach the level of 450 ppm Cl or even more. Such a level is considered damaging to sensitive crops such as citrus, avocado, and mango.

It seems sensible and rational to assume that the cost of treatment up to the level required by the rules aimed at the preservation of the environment will be borne by the producers of the waste—the urban population. Thus, the share to be borne by agriculture includes storage and conveyance. Under conditions of high profits from the agricultural use of treated wastewater, there might be a case for sharing the profits between farmers and municipalities (Dinar et al., 1986). However, such a situation seems unrealistic today. With all this taken into account, the range of costs of using treated wastewater in irrigation might be considered as falling between 10 to 25 cents per m^3.

Table 2.3. Cost of water from various sources (cents per m^3)

	Cost (cents/ m^3)	Percent
Natural sources	< 22	43
	22 - 40	32
	40 >	25
Treated wastewater	10 - 25	
Desalinated brackish water	40 - 50	
Desalinated seawater	80 - 120	

Table 2.3 presents a distribution of water supply sources according to the costs per m^3 of water. The cost of supply from natural sources is based on the Mekorot Work Plan and Budget (Mekorot Company, 1993). Mekorot is the national water company with most of the shares owned by the government. It is in charge of about two-thirds of the total water supply in Israel. The cheapest category of water from natural sources includes water supplied by private

non-Mekorot projects (Mekorot supplies about 65 percent of the total). Scrutiny of Table 2.3 suggests that the cost of treated wastewater is lower than about one-half of the water from natural sources, and it is considerably cheaper than the desalinization of brackish water by reverse osmosis and certainly than desalinization of seawater. Thus, treated wastewater can absorb the additional costs involved in upgrading quality and still be competitive.

5. THE USE OF WASTEWATER IN AGRICULTURE

There are two major interrelated problems involved in the strategy of using wastewater for irrigation: the interregional geographical allocation and the treatment level.

In general, there are few regions and a limited agricultural potential for low-level (secondary) treated wastewater. This situation will only be exacerbated due to the process of urbanization and increasing density of population.

A major concentration of wastewater sources prevails in the coastal plain where the urbanization process has been the greatest. However, this region is located above the coastal aquifer and there are considerable hazards with irrigation in this region even with very highly treated wastewater. Note that while the wastewater can be treated up to a level objectively equal to potable standards, uncertainty and lack of knowledge regarding carcinogenic materials and perhaps other currently unknown damaging elements are a source of worry and objection on behalf of those in charge of public health. On the other hand, there is an abundance of land in the South and the Negev that are not located above an unconfined aquifer, and secondary treatment level of effluent can be used there from the point of view of conserving groundwater.

There are two strategic alternatives widely discussed nowadays: (1) conveying most of the coastal plain wastewater south and shifting agriculture from the coastal plain to the Negev and (2) using high-quality effluent above the coastal aquifer. The latter alternative evades the currently prevailing restrictions. However, a recent document prepared by the researchers of the Volcani Institute of the Agricultural Research Organization (Fein et al., 1995) suggests that irrigation with treated wastewater in the coastal plain is environmentally feasible and at least not inferior to irrigation with quality of water from the National Water Carrier.

The advantages and disadvantages of the two alternatives are obvious. Conveying the wastewater to the Negev and expanding the agriculture there avoids the rigidities and restrictions derived from the currently prevailing system of land allocation and other institutional realities. The problem in the Negev is whether, under the current sociopolitical climate and the diminishing support of agriculture on behalf of Israeli society such a development can indeed take place. On the other hand, in the coastal plain, the land is perhaps the most fertile in Israel, with favorable climatic conditions, and there is still a viable core of good farmers with a strong infrastructure for farming. Furthermore, some parts of Israeli society place a high value on green and verdant open areas, despite the fact that they are giving way, gradually, to the forces of urban pressure.

6. TREATED WASTEWATER FOR ALL PURPOSES?

Under the current technology, highly treated wastewater meets all the objective standards of drinking water, yet dual-supply systems are developed in order to avoid any unforeseeable hazards. My own observations suggest that even triple systems are being introduced and expanded, such as the use bottled mineral water and home treatment of water.

It is proper to end this section with a question. Could highly treated wastewater serve as an emergency source for households under conditions of severe drought and severe water scarcity? Some water experts suggest constructing desalinization plants and use as an argument severe water scarcity situations with low probability of occurrence.

Could highly treated wastewater, supported by bottled drinking water, serve as an alternative solution?

7. WATER ALLOCATION AND PRICING

Fresh high-quality water is allocated to users according to an institutional quota system. These quotas were established in the early 1960s and have not been thoroughly revised since then. According to the Water Law of Israel and its underlying philosophy, the water is considered the property of the nation and is allocated for self-use purposes. Accordingly, quota transfers among users are illegal. Due to the developments since the early 1960s and the prevailing realities, transfers are practiced despite their illegality.

Various arrangements prevail regarding the allocation of brackish water and the right to use reclaimed wastewater. A major problem faced by the Water Commission is how to incorporate the low-quality water into the quota system and what should be the rate of substitution between high-quality and poor-quality water.

The water supplied by Mekorot to the farmers is priced according to a block differential (tier pricing) system: the first 50 percent of the quota has a low price (price A is 14.5 cents per m^3 in fall, 1994); the next 30 percent of the quota bears a higher price (price B is 17.5 cents); and the remaining 20 percent of the quota bears price C (23.5 cents in fall, 1994). Users exceeding the quota have to pay a considerably higher price with an element of penalty.

The pricing system enables interuser water mobility without changing the quota system and facing the struggle with farmers' rights to the quotas. It is also a venue for subsidizing farmers (via prices A and B), with potentially efficient prices at the margin of the quotas. Furthermore, no extra transaction costs are involved because all water supplied by Mekorot is metered and the charging system for water is automated.

This system is a mix of political and institutional allocations with the market mechanism being effective at the margin of the allocations. In other words, this is a mix of egalitarian and efficiency measures (see also Yaron, 1991). In the view of the author, market mechanism alone may lead to results incompatible with the noneconomic goals (national, social). Another advantage of the block differential pricing is that it takes away some of the rent, potentially accumulated by water suppliers, and transfers it to the farmers. Note that the marginal costs of water supply are increasing in shifting from one water plant to another within the same region. With the introduction of desalination, the shift in marginal costs of water supply may be substantial. For efficiency, the marginal water should be priced according to the marginal costs. But if all the water supplied is priced at the marginal cost, as free-market mechanisms suggest, a huge rent will be left in the hands of water supply companies. Block prices A and B leave some of this rent in the hands of the farmers without additional transaction costs.

While the current system has evident advantages, it is time for revision. First of all, the current quota system does not take into account the changes that have occurred since their establishment—either in terms of the growth of urban population, changes in agricultural production, the introduction of greenhouse technology, or the differential development of production systems of the farms, even in the same region.

In reality, the quotas are not observed. While the towns systematically exceed the quotas, the agricultural settlements use less water than is allocated to them by the quotas. Deviations from the quotas in recent years are presented in Table 2.4.

Thus the quotas should be revised. At the same time, complete cancellation of quotas to farmers is not feasible because farmers need some reassurance.

Another weakness of the current pricing system is that water is priced at the same level in all regions and the marginal price C does not necessarily reflect the real marginal cost of supply on a regional basis.

Table 2.4. Deviation from water quotas (%)

	Towns	Moshavim	Kibbutzim
1988	+17	-13	-11
1989	+18	-24	-17
1990	+28	-18	-11

8. URBAN USES

Water supplied by Mekorot to municipalities is charged 31 cents per m^3 (fall 1994 prices) for domestic use and 23.5 cents per m^3 for industrial use. However, the ultimate consumers pay considerably higher prices for water because the municipalities use the supply of water as a venue for taxation, thus making water supply a source of profit.

Table 2.5. Profit as a percentage of outlays in 1989 to 1990

City	Profit
Tel Aviv	44
Bat Yam	169
Herzeliyah	135
Ashdod	23
Raanana	80
Ramlah	6
All municipalities (simple average)	26

Research by Eckstein and Rosovski (1993)

Eckstein and Rosovsky (1993) observed that the richer the inhabitants of a community, the higher the percentage of profit from water supply (see Table 2.5).

9. SPECIAL ISSUES

9.1. Linkage in Supply Between Domestic and Agricultural Users

A considerable share of water supplied by Mekorot is delivered through plants that serve both the domestic and agricultural users. However, the parameters of demand by these two groups of users are in different (in terms of short- and long-term reliability of supply, water quality, and peak monthly average supply ratio).

As the water plants jointly supply domestic and agricultural users, the supply has to meet the most demanding parameter. Thus, in most cases, agriculture is served by plants that meet demand parameters not required by agriculture. The issue of cost-allocation between domestic and agricultural users is therefore raised. Qualitatively, the cost allocation policy and, accordingly, the pricing policy are clear. In order to arrive at the quantitative measure, research should be carried out. As the shortage of fresh water to both domestic and agricultural users becomes more severe in the years to come, the importance of this problem will become more acute.

9.2. Taxing Groundwater

In some regions of Israel, there are shallow aquifers with low pumping costs and therefore low water supply costs (for example, the Coastal Plain). There is a considerable difference between the private and the social costs of water in this region. The alternative for using this water in the region by private well-owners is conveying it to the Negev by the NWC and substituting for water from Lake Kinneret.

Until recently, the shallow aquifers' low-cost water was levied using the mechanism of an equalization fund. The fund that was collected was used to subsidize the national water system operated by Mekorot. Recently, the Water Commission has been looking for a more rational system. Taxing of groundwater is being discussed with economic parameters being used for setting up the tax level. The parameters under consideration include the distance from the Negev, water salinity, and the level of groundwater (a high level, possibly involving overflow to the sea, would be exempted from levy, whereas a low level of groundwater, which may lead to shortages in the following years, will be heavily levied).

9.3. Privatization

As previously mentioned, Mekorot Water Company controls 65 percent of the total water supply in Israel. The issue of privatization or partial privatization is being discussed.

This issue is only mentioned here, due to its importance, but due to its complexity, it falls beyond the scope of this chapter.

10. LIMITED ECONOMIC RESEARCH

As in many other countries, the water system in Israel is dominated by engineers, hydrologists, and agricultural experts. The economic research is relatively limited, while the demand for economic analysis is quite high, as evidenced by the previous examples.

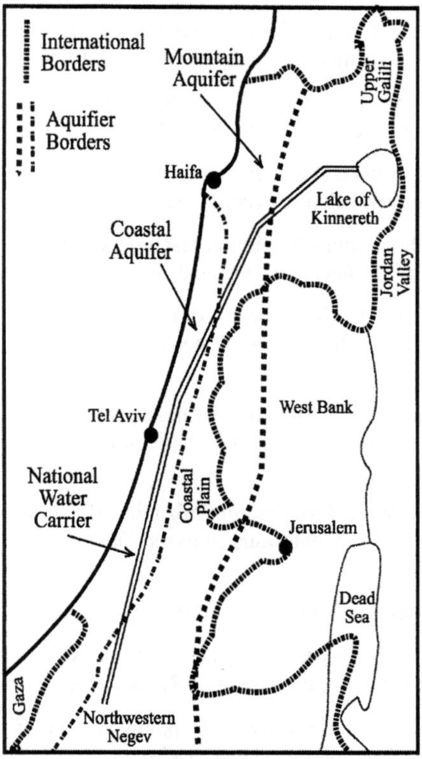

Figure 2.1. A schematical map of Israel and the core elements of its water system

11. SUMMARY

The chapter reviews the major issues of Israel water economy. The dominating feature is the increasing scarcity of high-quality water and the increasing share of low-quality water for agriculture. Water pricing methods and other economic measures aimed at increasing the efficiency of water use are discussed and evaluated.

REFERENCES

Avnimelech, Y. Use of wastewater in agriculture: position of ministry of quality of environment (Paper presented at the Third Continuing Workshop on Israel Water Issues). Center for Agricultural Economic Research. Rehovot, Israel. 1991. (Hebrew)

Dinar, A., D. Yaron and Y. Kannai. Sharing regional cooperative gains from reusing effluent for irrigations. *Water Resource Research,* 22: 339-344. 1986.

Eckstein, S. and S. Rosovski. Water economy in municipalities (Research Report submitted to Water Commission) Bar-Ilan University. Ramat Gan, Israel. 1993. (Hebrew)

Falkovitz, M. and E. Feinerman. Minimum leaching scheduling of nitrogen fertilization and irrigation. *Bulletin of Mathematical Biology,* 56:665-686. 1994.

Fein, P., N. Haruvy and Y. Schainberg. Ra'anana wastewater treatment project: criteria for quality under different alternatives. Water and Soils Institute, Volcani Center. Bet Dagan, Israel. 1995. (Hebrew)

Feinerman, E. and A. Dinar. Economic and managerial aspects of irrigation with saline water (Working paper no. 9103). Center for Agricultural Economics Research. Rehovot, Israel. 1991.

Feinerman. E. and H. Voet. Dynamic optimization of nitrogen fertilization of citrus and the value of information from leaf tissue analysis. *European Review of Agricultural Economic,s* 22:103-118. 1995.

Mekorot Company. Mekorot Work Plan and Budget (Mimeo). Tel Aviv, Israel. 1993. (Hebrew)

Nevo, N. Comments (Paper presented at the Seventh Continuing Workshop on Israel Water Issues). Center for Agricultural Economics Research. Rehovot, Israel. 1992. (Hebrew).

Shevah, Y. and A. Shelef. Wastewater 2000: policy and plan for treatment and reuse of wastewater. Israel Water Commission. 1993. (Hebrew)

Schwartz, Y. Israel Water Sector Study (Unpublished report prepared for the World Bank) Ben Gurion University. Beer-Sheva, Israel. 1990.

Tahal Consulting Company. Water Master Plan (Interim report). Tel Aviv, Israel. 1998. (Hebrew)

Yaron, D. Allocation of Water and Water Prices in Israel. *Economic Quarterly,* 150:465-478. 1991. (Hebrew)

Yaron, D. Economic aspects of irrigation with low quality (saline) in: "State of the Art: Irrigation, Drainage and Flood Control No. 3" (K. K. Framji, ed.). 263-286. International Commission on Irrigation and Drainage. New Delhi, India. 1984.

Yaron, D. An approach to the problem of water allocation to Israel and the Palestinian entity. *Resource and Energy Economics,* 16:271-286. 1994.

3

EMPIRICAL ANALYSIS OF THE DEMAND FOR WATER BY ISRAELI AGRICULTURE*

Dan Yaron
Harvard University, Boston, Massachusetts, USA

1. INTRODUCTION

The aim of this chapter is to review the methodology followed by an empirical analysis of the demand for water by Israeli agriculture (Yaron, 1966) and to stress the thesis—basic to the approach—that in the economic analysis of the complex of problems of the water resource in agriculture, irrigation water can and should be referred to as any other production factor which takes part in the process of agricultural production. Concerning the demand for irrigation water, our approach refers to water within the framework of production functions, from which the demand for water is derivable. On the supply side, water is referred to as an output producible at variable, man-controlled output levels. The study concentrates on the analysis of the agricultural demand, utilizing results of analyses regarding supply and urban demand performed by experts in the other fields.

Two lines of attack were followed: (1) normative and positive studies on the farm level in two selected regions aimed to analyze in considerable detail the microstructure of the demand for water of farms of various types; (2) a normative analysis on a countrywide regional level aimed to derive an approximation of the demand function for water by Israeli agriculture, in relation to her agricultural development program (Israel Ministry of Agriculture, 1965).

Considering the whole hierarchy of product-factor demand interrelationships in the field of water resources, the following seems to be a logical succession of elements in this complex: (a) the water production (or response)

---

*Permission to publish this chapter was granted by the American Agricultural Economists Association. The chapter was originally published under a similar title in *The Journal of Farm Economics,* 49:461-473, 1967, while Dan Yaron was on leave from The Hebrew University.

function of a single crop; (b) the production function of an individual farm, with water as one of the production factors; (c) the demand for water of an individual farm as derived from its production function; (d) the demand for water of a group of farms, of an agricultural region, or of any higher complex of interrelated regions; (e) the sociopolitical economic mix, which determines the framework within which the actual demand for water is derived.

In the following, we review the above elements as they were reflected in the empirical analysis. A review approach has been chosen for the presentation, thus focusing on the general framework of the study rather than on particular details.

2. THE WATER PRODUCTION FUNCTION OF A SINGLE CROP

One of the major current theories[1] regarding water-soil-plant relationship contends that plants respond differentially to soil moisture content and that changes in the moisture regime during the period of plant growth result in corresponding changes in yields of the irrigated crops. Accordingly, it is appropriate to refer to the relationship between the input of irrigation water and the yield of a crop within the conceptual framework of a production function. Studies, which adopted this view, have been recently reported by Beringer (1961), Dorfman (1963), Moore (1961), and others.

The major problem of empirical estimation of such functions is their specification, and in particular the choice of the independent variables, the dependent one being the crop yield per land unit. Different approaches to this question have been suggested (Beringer, 1961; Dorfman, 1963; Moore, 1961). The approach chosen in our study was to relate the yield to the fundamental irrigation variables, which can be controlled by the farmer and are, therefore, operationally meaningful. The following general specification was applied:

$$y = f(x_1, x_2, x_3)$$

where

y = crop yield per land unit area,
x_1 = depth of soil moistening,
x_2 = total quantity of water applied, and
x_3 = frequency of irrigation (or number of irrigations).

It should be noted, however, that in nearly all of the 29 irrigation experiments analyzed, a high correlation was found between x_2 and x_3, and therefore the function was usually reduced to $y = f(x_1, x_2)$. If the empirical data allow for a more detailed specification, subdivision of the overall irrigation period into several subperiods should be useful for gaining additional information. In such a case, the variables x_l would be discriminated according to a subperiod index. None of the experiments with various field crops analyzed by the author and collaborators (Oron, 1965; Subotnik, 1964; Yaron, 1966) allowed for such discrimination.

Table 3.1. Least-squares estimate of y (sorghum fiber yield kg/dunam) as a function of x_2 (effective quantity of water applied m³/dunam[a]), using two alternative formula[b]

Year	Fitted formulas for y	R^2
	Equation (S-1): $y = b_0 + b_1x_2 + b_2x_2^2$	
1958	$5.56 + 2.45x_2^{**} - 0.00207x_2^{2\,**}$	0.95
1959	$229.9 + 2.602x_2^{**} - 0.0024x_2^{2\,**}$	0.96
1961	$142.1 + 2.45x_2^{**} - 0.00239x_2^{2\,**}$	0.94
1958-61[c] (combined)	$124.76 + 2.47x_2^{**} - 0.00217x_2^{2\,**}$	0.96
	Equation (S-2): $y = b_0 + b_1x_2 + b_2x_2^3$	
1958	$14.67 + 2.095x_2^{**} - 0.0000025x_2^{3\,**}$	0.96
1959	$248.12 + 2.033x_2^{**} - 0.0000026x_2^{3\,**}$	0.95
1961	$147.75 + 2.091x_2^{**} - 0.0000035x_2^{3\,**}$	0.95
1958-61[c] (combined)	$137.9 + 2.011x_2^{**} - 0.0000025x_2^{3\,**}$	0.96

**Significant at the 1 percent level

[a] 1 dunam = 0.1 hectare

[b] Row spacing 75 cm. Irrigation depth was fixed (1.8 m) throughout the various quantities of water applied.

[c] Estimated according to the model:

$$y_{ikp} = b + a_i + b_1z_{1ik} + b_2z_{2ik} + \dots + e_{ikp}$$

where

y_{ikp} = yield obtained in *i*th year from *p*th replicate of *k*th treatment;

a_i = effect of *i*th year;

z_{j1k} = the level of *k*th treatment in *i*th year, of the factor corresponding to the regression coefficient b_j; and

e_{jkp} = random error term, including the block (replicate) effect

For the sake of illustration we present, in Table 3.1, empirical estimates of the water production function in sorghum, using two alternative algebraic forms.[2] The corresponding curves are shown in Figures 3.1 and 3.2.

The estimates presented in this illustration, as well as the analyses of other irrigation experiments suggest that the curves fitted for a given crop in the same location but different years tend to run parallel to each other, in the main discrepancy being in their overall elevation. The difference in elevation may be explained in terms of differences in soil fertility (between the experimental fields) and other effects specific to the particular years (such as weather, pest infestation, and the like). This observation leads to the formulation of a hypothesis that the "year effect" is of an additive nature and does not affect the slopes of the curves in the particular years, namely that under conditions at one and the same location, the marginal yield of a crop is a function only of the quantity of water (other irrigation variables being equal). The verity of this hypothesis is essential for the use of empirical estimates of water response functions in predicting the effect of controlled variations in the quantity of water applied, on the yield of irrigation crops.

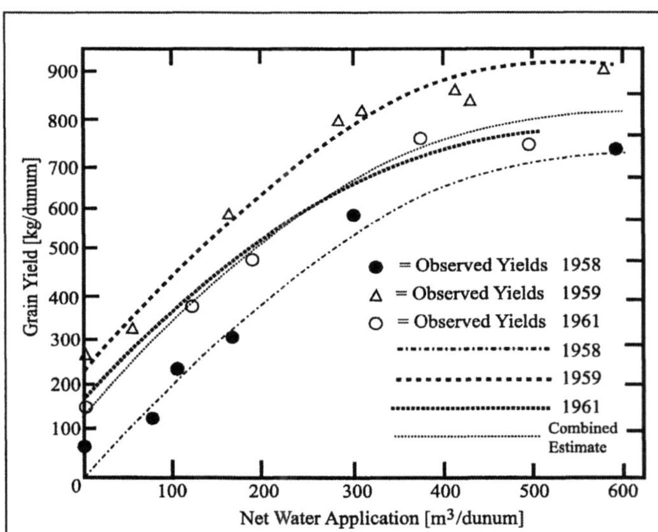

Figure 3.1. Estimated relationship between water quantity and sorghum grain yield, using the formula $y = b_0 + b_1x_2 + b_2x_2^2$

So far, the results of empirical analyses of 29 irrigation experiments suggest acceptance of the above as a working hypothesis. However, further empirical evidence is needed for its support and for a better understanding of the nature of water response functions. Of particular importance is the question of generalization of results and guides for extrapolation from one location to another. The importance of such extrapolations is emphasized by the fact that irrigation experiments are costly and no single region has adequate research funds to be self-sufficient in this respect. Towards this goal, it seems useful to have an interregional and international pool of information on irrigation experiments. With sufficient amounts of data so gathered, it might be possible to incorporate in the analysis variables, which characterize soil and weather conditions and be able to arrive at interlocation interpolations.

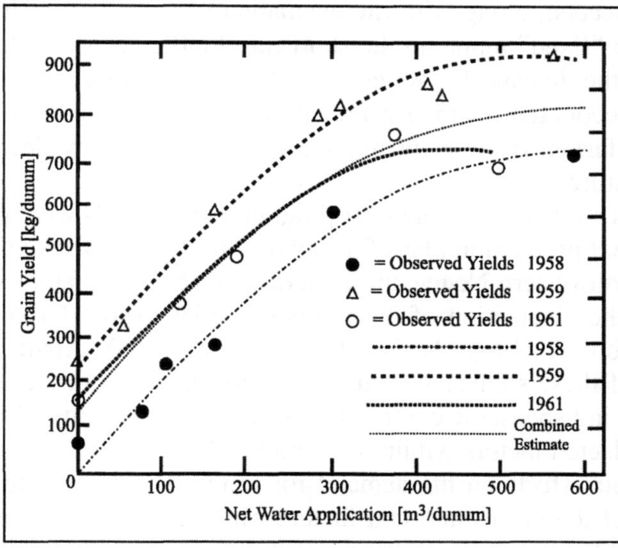

Figure 3.2. Estimated relationship between water quantity and sorghum grain yield, using the formula $y = b_0 + b_1x_2 + b_2x_2^3$

3. THE AGGREGATE PRODUCTION FUNCTION AND THE DEMAND FOR WATER BY A SINGLE FARM

The response functions to water in individual crops constitute the basic elements for the derivation of the aggregate production function of the farm, the derivation being subject to efficiency criteria. If the response functions are not known, the derivation of the aggregate production function of the farm is restricted to rigid water inputs of the individual crops ("irrigation norms," which represent particular points on the individual response surfaces). Obviously, in the second case, the farm's output for a given bundle of resources cannot be higher than that obtainable under conditions of complete information.

A farm's demand function for water is directly derivable from its production function. Two facts, however, should be kept in mind in this context. First, recall that such derivation is based on the assumption of rational behavior of the producers. The validity of this assumption seems to be the major dilemma involved in the transition from theory to empirical estimates of the demand for water, based on the production function approach. Fortunately, it seems that in Israel farmers do indeed tend to behave rationally, in the economic sense.

Second, even if farmers approximate economic rationality, the phenomenon of inefficient application of certain "fixed" inputs, reflected in improper proportions with other inputs, is not uncommon in the short run. Therefore, discrimination between a succession of short run and long run demands for water is needed, along with the evaluation of the effect of disequilibrium in any of the "fixed" inputs in the short run. Inefficiencies of this sort and their effect on the demand for water, as well as the prospects of their elimination, must be accounted for in any attempt to estimate empirically the demand for water by farms; if the estimates are to serve for policy decisions extending into the future.

In this study numerous linear programming models were performed to derive optimal production plans for kibbutz and family farms in two selected regions (northwestern Negev and eastern Jezreel Valley) at varying amounts of water at the disposition of the farms (Frumkin, 1965; Oron, 1965; Shapiro, 1966; Subotnik, 1964; Weiss, 1965, Yaron, 1966). Apart from the major product of these studies—normative estimates of the farms' demand for water—their indispensable contribution lies in the possibility of controlled variation of selected factors within the framework of the planning models and analyzing their effects on the demand for water. The studies performed suggest that two of the most important factors which affect the demand for water are the particular irrigation techniques applied on the farms and the degree of mobility of farm labor, as reflected in the alternative cost of labor. To illustrate, considering some farming situations in an analysis of a model of family farm in the Negev area, the incorporation of flexible irrigation practices for major field crops, formulated on the basis of water response functions of these crops, raised the marginal productivity of water by 15-30 percent, and in one case by 50 percent, as compared with the conventional irrigation practices restricted to rigid "norms." The same analysis suggests that the difference between marginal productivity of water on family farms in the area with no *a priori* alternative cost attached to their labor can be 30-50 percent higher than on similar farms but with an alternative cost of labor at the predominating wage in the region. On the other hand, in programming analyses of kibbutz farms in the same region, it was found that variations in the alternative cost of labor had only a minor effect on the marginal productivity and the demand for water. This difference is due to the comparatively low labor/land and labor/water input ratios on kibbutz farms as compared with family farms. The above results have been presented here as an illustration, in order to emphasize the argument that the analysis of the demand for water by a farm or group of farms, cannot be performed satisfactorily without a comprehensive study of the production patterns of the farms, their trends of development, and the like. Another observation of interest, which emerges from the micro-farm studies is that since the relatively profitable farm products are subject to production quotas, the marginal (and the relevant) segment of the farm's demand function for water is determined by a relatively small group of products.

On a series of kibbutz farms in Israel, programming analyses of this kind were performed in parallel to the preparation of the yearly production plans of the farms and in cooperation with the farms' officers (Frumkin, 1965; Shapiro, 1966; Weiss, 1965). Thus the estimates of these farms' demand for water are derived from the actual production plans and are realistic to the extent that the production plans materialize.[3]

4. THE DEMAND FUNCTION FOR WATER OF A GROUP OF FARMS OR AN AGRICULTURAL REGION

The major difficulty which arises in shifting from the demand for water of individual farms to the aggregate demand of a group of farms or a region is the question of which aggregation pattern should be applied. The horizontal aggregation rule applies only to a situation in which the interfarm water allocation follows the equimarginal principle. However, the common situation in many countries is that water allocation is highly influenced by institutional, social, and political factors, as a result of which the economic efficiency of interfarm water allocation is often distorted. Under such circumstances, elaboration of the aggregation rules is needed. Three major situations can be distinguished: a) Interfarm water allocation is determined without any consideration of allocative efficiency. It is obvious that, in this case, the concept of economic demand for water becomes meaningless. b) The other extreme is a completely efficient allocation, in which case the horizontal aggregation is applicable. c) The third situation is one of an intermediate nature, in which the institutional dominance in water allocation is restricted to the currently available water resources, while the concept of *economic* demand is related to additional supply. This implies a subdivision of the water market into two segments, the first one being wholly institutional and the second having a competitive nature. Such an approach has some appeal to policy makers who, on the one hand, are reluctant to interfere with the historically established water allocation system, but on the other hand, would refrain from promoting development of new water projects unless they are economically justified. The allocation implication of this approach is that newly developed supplies should be allocated first to farms with the highest marginal productivity of water. At the same time farms will remain with institutionally appropriated water rights and marginal productivity of water lower than the marginal cost of supply. A demand function for water derived through assumptions of this nature has been requested by some of the agricultural institutions and farm leaders in Israel. If such an approach is followed, we may consider the institutional water allotments as predetermined and refer to the demand for water over and above this quantity, within a framework of efficiency, with its demand aggregation implications.

Note that situation *c* may be regarded as a subcase of *b* with certain constraints upon the allocation system. Actually numerous variants of situation *c* may be visualized as the number of constraints or their levels vary. To summarize the above discussion, it may be said that whether situation *b* or *c* is the relevant one, the aggregation of the individual demands to the demand function of the group follows the same rules. The only difference is that, in situation *c,* a given quantity of water is considered as preallocated, beyond the realm of efficiency considerations.

It should be noted that the distinction between "efficient" and "inefficient" interfarm water allocation is not so clear-cut as it may seem in view of the above discussion. Consider, for example, a backward farming region, A, in which the marginal value product of water is equal to that of adjacent region B, with which region A competes for limited water supplies. (Assume that transfer cost of water is negligible.) Suppose that the government intends to raise the farm income in region A, and implements its policy by influx of capital to farms of region A (say through loans on easy terms aimed to raise the productive capacity and income of the farms). The additional capital may raise the marginal productivity of water in region A, as the result of which water should be transferred from region B to region A. Such a transfer will normally be designated as "efficient;" however, if the influx of capital to region A was not in accordance with efficiency criteria, the overall economic efficiency of this step may be questioned. This amounts to saying that the concept of economic efficiency may be misleading if referred to in a partial framework.

5. THE DEMAND FOR WATER OF INTERRELATED REGIONS

The aggregation considerations of the previous section apply as well to the derivation of the demand for water of higher groups in the farm-region-nation hierarchy. The only additional feature is that the cost of conveying water is now generally non-negligible and should be taken into account in the formulation of the conditions of efficient allocation. Spatial equilibrium or interregional competition models are apt to provide a tool for the analysis of optimal interregional allocation (or, rather, optimal under constraints) and the derived demand for water in a normative framework (Heady and Egbert, 1959; Takayama and Judge, 1964 a and b; Yaron et al., 1965). Such a model has been applied to this study and the analysis of the agricultural demand for water on the countrywide level. The programming analysis has been restricted to the marginal segment of the demand function, with the premarginal segment being evaluated on the basis of the agricultural development program.[4] In the programming model, water constitutes one of the limited resources for which the regions compete, the overall supply of water and the interregional conveying system being predetermined. The results indicate the optimal allocation of marginal quantities of water and the regional pattern of the marginal productivity or the demand price for water. By varying parametrically the assumptions with respect to the supply of water, additional points on the regional demand functions for water were derived, the demand functions of the various regions being linked (under the assumption of economic efficiency) by price differentials equivalent to cost of interregional transfer of water. Assumptions with respect to the efficiency of water allocation and the allocation of certain products have been varied in several variants of the model and their effect on the demand for water analyzed.

The problems involved in such a line of analysis may be divided into two major categories: conceptual and technical. Let us start with what we consider the major conceptual problem of interregional programming models. The essence of programming is to maximize a given objective function subject to a given set of constraints. It is probably not wrong to state that the higher the programming level in the farm-region-nation hierarchy is, the more difficult it becomes to formulate a meaningful objective function of the programming model. This results from the multiplicity of goals of the society and the difficulties involved in their explicit formulation and in assigning relative weights to the various goals.

Considering a situation in which society is confronted with *n* different goals, without specified rates of substitution between these goals, one operational approach would be to derive (through programming) the efficiency frontier in this *n*-goal space, as a basis for policy decisions. Apparently, it would be easier for policy makers to choose a given point on the boundary of this feasibility set rather than, a *priori,* to assign relative weights to the various goals, necessary for formulation of the "overall objective function" of the society. For example, considering development of water resources for agriculture, the two major and competing goals seem to be allocative efficiency and income redistribution. It is apparently easier for any policy decision to rank the preferences with respect to the feasible combinations of efficiency and income distribution, rather than to formulate *a priori* the pattern of preferences over the whole space of these two goals.

However, the limitations of an empirical study often impose restrictions on the number of alternative formulations of the model needed for the derivation of the efficiency frontier in the society goals space. The analyst has to choose only one of several formulations, the choice being made with the aim of approximating reality as much as possible. In a democracy, the "overall decision set" of the society regarding current agricultural production and agricultural development programs is determined through the interaction of government policy and farmers' objectives, with each farmer striving to maximize his own income. While the ultimate decision on what to produce is up to the farmers, the government has at its disposal numerous instruments for the control of the economic activity of the producers. If we take the position that it is much easier to approximate the objectives of the farmers (the predominating one being profit maximization) and to express them in the objective function of the interregional programming model, than to express quantitatively the goals of the higher hierarchies (e.g., income redistribution), it follows that the appropriate set-up of the model is formulating the objective function in terms of maximization of farmers' income, subject to the following restrictions: (a) competitive behavior of the farmers; (b) available resources and technology; and (c) restraints imposed by other goals of the society as expressed in government decisions. Interregional competition models meet the above specification.

It should be noted that any effective constraint, incorporated into the model, which originates in one of the goals other than maximization of farmers' income within a competitive structure, may (a) reduce the value of the overall objective function attainable without this restriction, and (b) affect the shadow prices of the restricted resources and the demand for these resources. This, of course, also holds for irrigation water, if water is regarded in the model as a restricted resource. That is, any binding policy restriction imposed on the program may distort the marginal productivity and the demand for water. The direction of such a distortion is hard to judge *a priori* without a detailed analysis. Note also that policy measures must not necessarily be expressed in terms of restrictions; they may be expressed in planning coefficients or otherwise. Whatever various policy measures are incorporated into the analytical model, their effect on the demand for water should be duly accounted for. In particular, it is of interest to evaluate the long-run effects of the current policy measures and the prospects of their continued implementation in the future, as well as their implications for water resource development. It is at this stage that the problem escalates to the highest of its levels, namely the interrelationship between the economic and sociopolitical factors and their combined effect on the demand for water.

Two major problems of a technical nature faced by the study were (a) interfarm incorporation into the model of demand functions for agricultural products (from which the demand for water is derived), and (b) the actual formulation of the model in order to make it both meaningful and manageable. Regarding issue (a), appropriate programming algorithms were developed, varying with respect to the types of the demand functions incorporated. All algorithms lead to the derivation of the following competitive equilibrium conditions:

(1) $A^t X^t \leq B^t,$

(2) $\sum_{t=1}^{T} D^t X^t \leq B^*,$ $\qquad t = 1, 2, \ldots, T$

(3) $A_j^{t\prime} U^t + D_j^{t\prime} U^* \geq c_j^t$ $\qquad$ for any product j with fixed prices, $j = 1, 2, \ldots, n$

or

(4) $$A_j^{t'}U^t + D_j^{t'}U^* + h_j^t \geq \phi_j\left[\sum_{t=1}^{T} x_j^t, \sum_{t=1}^{T} x_s^t\right]$$

for any product j with varying prices,
$s = 1, 2, \ldots, n, \quad s \neq j$

and

(5) $$X_t U \geq 0,$$

where

A^t = matrix of input coefficients of limited resources specific to producer t,

$A_i^{t'}$ = vector of inputs of limited resources specific to producer t, per unit of activity j;

X^t = vector of activity levels of producer t;

x_j^t = level of activity j of producer t;

B^t = vector of limited resources specific to producer t;

D^t = producer's t matrix of input coefficient of limited resources "common" to all (or some) producers for which the producers compete (irrigation water being one of these resources);

$D_j^{t'}$ = producer's t vector of inputs of limited resources "common" to all (or some) producers, per unit of activity j;

$B*$ = vector of the limited "common" resources;

U^t = vector of shadow prices of the limited "common" resources;

c_j^t = net income per unit of activity j (with fixed prices);

h_j^t = pecuniary cost of producer t per unit of activity j with varying prices, and

ϕ_j = price of product j expressed as a function of the total output of product j $\left(\sum_t x_j^t\right)$ and the total output of the related products.

$$\text{s}\left(\sum_t x_s^t,\ s = 1,2,..,n,\ s \neq j\right)$$

The above conditions may be regarded as the essence of any competitive equilibrium problem (Lee, 1958). Regarding them as the starting point, the related non-linear programming problem should be formulated so that its optimal solution will satisfy (1) through (5). The details of the developed algorithms are presented elsewhere (Yaron, 1966; Yaron et al., 1965).

As in any interregional programming study, the size of the model and its practical manageability were a major issue. Accordingly, several practical shortcuts were applied, which may all be classified under the heading of incorporation of mathematical programming and "naive" procedures.

Because of space limitations and the fact that these shortcuts were primarily related to the specific conditions of Israel, they are not detailed here.[5] At the same time, it should be stressed that the search for shortcuts, which arises from the particular nature of the problem seems to be a very important part of any study of this sort.

6. CONCLUSION

In this chapter, the approach applied to the analysis of the demand for water by Israeli agriculture is presented. The core of the study was an aggregative countrywide interregional programming model, with the aid of which the marginal segment of the demand function for water was derived, as a basis for the evaluation of the overall demand function. The interregional analysis was supported on the one hand by a series of analyses on the farm level, aimed at the exploration of the microstructure of the demand for water, and, on the other hand, by projections and programs regarding the overall development of Israeli agriculture.

The results of the study suggest that the shape of the demand function for water is highly dependent on the sociopolitical economic mix, which determines the framework within which it is derived. In particular, of substantial effect is the degree of efficiency in water allocation, which dictates the rules of aggregation of the demands of individual farms into the demands of higher groups. Therefore, the more comprehensive and realistic the agricultural development program or projection available as a background for the derivation of the agricultural demand function for water, the more realistic its estimate may be.

REFERENCES

Beringer, C. An economic model for determining the production function for water in agriculture. Giannini Foundation Report. Berkeley, California. 1961.

Bielorai, H. and J. Rubin. A study of the irrigation requirements and consumption water use of sugar beet in the northern Negev. *Agricultural Research Station Special Bulletin,* 6. Rehovot, Israel. 1957.

Bielorai, H. and D. Shimshi. The influence of depth of wetting and the shortening of the irrigation season on the water consumption and yield of irrigated cotton. *Israel Journal of Agricultural Reearch,* 13:55-62. 1963.

Dorfman, R. Response of agricultural yields to water in the former Punjab. The White House panel on water logging and salinity in West Pakistan report on land and water development in the Indus Plain. Washington, DC. 417-435. 1963.

Frumkin. D. Optimal water allocation on a kibbutz farm in the Gilboa region (Unpublished MSc. Thesis). The Hebrew University of Jerusalem. Rehovot, Israel. 1965. (Hebrew)

Furr, J. R. and C. A. Taylor. Growth of lemon fruits in relation to moisture content of the soil. *United States Department of Agriculture Technical Bulletin,* 640. 1939.

Hagan, R., M., Y. Vaadia and M. S. Russel. Interpretation of plant responses to soil moisture regimes in: "Advances in Agronomy" (A. G. Norman, ed.). New York Academic Press, 11. 1959.

Hamilton, J., C. O. Stanberry and W. M. Wooton. Cotton growth and production as affected by moisture, nitrogen and plant spacing on the Yuma Mesa in: "Proceedings of the Soil Science Society of America," 20:246-252. 1956.

Heady, E. O. and A. C. Egbert. Programming regional adjustment in grain production to eliminate surpluses. *Journal of Farm Economics,* 41:718-733. November 1959.

Israel Ministry of Agriculture. Agricultural Planning Center. Preliminary production plan for agriculture for 1966/67-1970/71 (Mimeo). Tel Aviv, Haqirya, Israel. 1965. (Hebrew)

Lee, I. M. Optimum water resource development: preliminary statement of methodology for quantitative analysis. Giannini Foundation Report, 206. Berkeley, California. 1958.

Moore, C. V. A General analytical framework for estimating the production function for crops using irrigation water. *Journal of Farm Economics,* 43:876-888. November 1961.

Oron (Simchah), A. Marginal productivity and demand for water of moshav farms in the Taanach area (MSc. Thesis). The Hebrew University of Jerusalem. Rehovot, Israel. 1965. (Hebrew)

Shapiro, A. Study in process toward the MSc. Degree. The Hebrew University of Jerusalem. Rehovot, Israel. 1966.

Subotnik, A. Estimates of response functions to irrigation water in selected crops and their incorporation in programming of a moshav farm in the northwestern Negev (Unpublished MSc. Thesis). The Hebrew University of Jerusalem. Rehovot, Israel. 1964. (Hebrew)

Takayama, T. and G. G. Judge. Spatial equilibrium and quadratic programming. *Journal of Farm Economics,* 46:67-93. February 1964a.

Takayama, T. and G. G. Judge. An Interregional activity analysis model for the agricultural sector. *Journal of Farm Economics,* 46:349-635. May 1964b.

Weiss, M. Optimal production programs and estimates of marginal productivity of water on kibbutz farms in the northwestern Negev (Unpublished MSc. Thesis). The Hebrew University of Jerusalem. Rehovot Israel. 1965. (Hebrew)

Yaron, D. Economic criteria for water resource development and allocation. Department of Agricultural Economics. The Hebrew University of Jerusalem. Rehovot Israel. 1966.

Yaron, D., Y. Plessner and E. O. Heady. Competitive equilibrium—application of mathematical programming. *Canadian Journal of Agricultural Economics,* 13:65-79. 1965.

---

[1] This theory, advocated by Furr and Taylor (1939), Hagan et al. (1959), Hamilton et al., 1956, and others, is supported by numerous irrigation experiments (Bielorai and Rubin, 1967; Bielorai and Smimshi, 1963; Israel Ministry of Agriculture, 1965).

[2] The estimates were derived on the basis of irrigation experiments at the Gilat Experiment Farm in southern Israel in 1958, 1959, and 1961 (Subotnik, 1964; Yaron, 1966, Supplement A).

[3] As an "antidote" to the normative studies positive estimates of the marginal productivity of water were derived for farm samples in the same areas, through Cobb Douglas production functions, and residual value imputation techniques.

[4] This was possible because the demand for the marginal water in Israeli agriculture originates from a distinguishable group of products, namely irrigated field crops.

[5] For details see the complete report of the study (Yaron, 1966, pp. 157-164).

4

WHEAT RESPONSE TO SOIL MOISTURE AND THE OPTIMAL IRRIGATION POLICY UNDER CONDITIONS OF UNSTABLE RAINFALL*

Dan Yaron and Gadi Strateener
The Hebrew University of Jerusalem, Rehovot, Israel
The Center for Agricultural Economic Research, Rehovot, Israel

Dani Shimshi
The Volcani Center of Agricultural Research, Bet Dagan, Israel

Mordechai Weisbrod
The Hebrew University, Jerusalem, Israel
The Center for Agricultural Economic Research, Rehovot, Israel

1. INTRODUCTION

An approach is presented to estimating a response function of wheat yield to soil moisture and to determining the optimal irrigation policy under conditions of stochastic rainfall. A necessary condition for the determination of an optimal irrigation policy under unstable rainfall conditions is the availability of information on variation of the soil moisture over time and as a function of depth. Since it is impractical from the cost point of view to conduct measurements of soil moisture before and after each rainfall and/or irrigation, a method designed to reconstruct the soil moisture fluctuations on the basis of the incomplete data available is needed. A computer simulation model designed for this purpose is briefly described in the first section of this paper, and estimates of wheat response functions to soil moisture are given in the

*Permission to publish this chapter was granted by the American Geophysical Union. The chapter was originally published under a similar title in *Water Resources Research,* 9(5):1145-1154, 1973.

second section. An analysis of the optimal irrigation policy is presented in the third section.

Considerable data are available on methods aimed at evaluating evapotranspiration and fluctuations in soil moisture (e.g., Penman, 1949; Blaney and Criddle, 1950; Thornthwaite and Mather, 1955; Shaw, 1964; Jensen, 1967). Methodological expositions and empirical estimates of crop response functions to irrigation water or soil moisture have also been made (e.g., Moore, 1961; Dorfman, 1963; Yaron et al., 1963; Flinn and Musgrave; 1967; Baier and Robertson, 1967; Hall and Butcher, 1968; Yaron, 1971). Burt and Stauber (1971), among others, have analyzed optimal irrigation policy within a control-oriented conceptual framework. However, only a few studies that analyze the whole complex as an integrated system (e.g., de Lucia. 1969; Dudley et al., 1971a and b) have been published. We now attempt to provide such an analysis, i.e., an integrated-system approach to the overall complex of soil moisture-crop response function and optimal irrigation policy.

The empirical data that provided the basis for this study were obtained from an experiment on wheat irrigation of varieties FA and 46 performed by D. Shimshi at the Gilat Experiment Station in the Negev region, Israel, during four seasons (1965/1966-1968/1969). Some details of these experiments are presented in Table 4.1.

2. MODEL FOR TRACING FLUCTUATIONS OF SOIL MOISTURE OVER TIME AND DEPTH

The processes of water infiltration into the soil profile and of evapotranspiration are affected by the hydraulic properties of the soil, the type of vegetation, the quantity and frequency of rainfall and/or irrigation water, and the evaporative conditions.

A schematic view of the variation of moisture over time in a given soil layer is shown in Figure 4.1. The curve indicates the average moisture in the layer (moisture variability within layers is neglected). The following assumptions were made in drawing the fluctuation curve of Figure 4.1: (1) The maximal moisture content that the soil can- retain is the field capacity (FC) percentage, and the minimal content below which the soil moisture cannot fall is the permanent wilting point (PWP). The soil moisture can fluctuate only between these two levels. (2) Between irrigations (and/or rainfall events) the moisture content is continuously reduced by the process of evapotranspiration. (3) Under conditions of adequate water supply from irrigation or rainfall to a given soil layer with moisture content below FC, the layer will first achieve the FC status, and any water in excess of this amount will be drained to a lower layer. The process of water infiltration and drainage are assumed to be completed by the end of the day during which irrigation has been applied. (see Weisbrod et al., 1971). Here only the essentials are repeated.

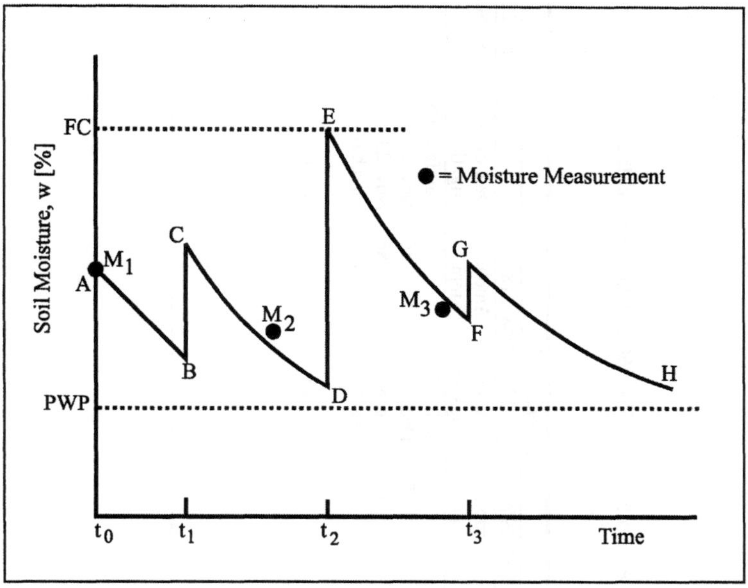

Figure 4.1. Typical moisture fluctuation in a given soil layer

The moisture fluctuation curve for a given layer starts at point M_1 (time t_0), for which moisture measurements are available. During the period $\Delta t = t_1 - t_0$ there was neither rain nor irrigation, and the process of evapotranspiration brought the soil moisture to point *B* at time t_1. At t_1 water was added to the soil layer, raising the moisture content to point *C,* and the process of evapotranspiration again started. At time t_2 a new quantity of water was added to the soil, raising the soil moisture to FC, the evaporation process started again, and so on.

It was assumed that during the whole period covered in Figure 4.1 there were only three moisture measurements, marked by points M_1, M_2, and M_3. Point *A* marks the initial soil moisture. In addition, the timing and quantity of rainfall and irrigation water, based on assumptions 1-3 above, are given. Under these circumstances the goal is to estimate and reconstruct the discontinuous curve, *AH,* in a way that 'best fits' the moisture measurements following M_1 (i.e. M_2 and M_3).

Table 4.1. Details on irrigation treatments of wheat in 4 years of experimentation, 1965-1966 to 1968-1969, at the Gilat Experiment Station

Treatment No.	Rainfall, mm[a]							
	1965-1966: 186		1966-1967: 297		1967-1968: 318		1968-1969: 204	
	Date of Irrigation	Amount, mm	Date of Irrigation	Amount, mm	Date of Irrigation	Amount, mm	Date of Irrigation	Amount, mm
1	Rainfall only		Rainfall only		Rainfall only		Rainfall only	
2	Nov. 26	150	Nov. 28	150	Dec. 8	86	Nov. 29	150
3	Nov. 26	150	Nov. 28	150	Dec. 8	86	Nov. 29	150
	Feb. 3	68	April 4	91	March 18	138	March 6	150
	March 10	130						
4	Jan. 24	100	Dec. 13	40	Feb. 26	90	Jan.16	144
							March 17	145
5	Jan. 5	100	Dec. 18	150	Feb. 25	90	Jan. 26	150
					April 10	110		
6	Jan. 5	100	April 9	120	March 14	151	March 2	162
	March 10	130						

[a] mm = millimeter

It was assumed that the evapotranspiration process is subject to the following general law:

$$(1) \qquad ET = dw/dt = a + bw \qquad a < 0, \quad b > 0$$

where w is the daily mean soil moisture content of the layer, measured by weight percentage; ET is the daily evapotranspiration from a given soil layer in terms of change of weight percentage of moisture, equal to $-dw/dt$; and a, b are parameters, depending on the soil type.

The function (1) is a linear differential equation that can be written in a separable form as

$$(2) \qquad dw/(a + bw) = -dt$$

with the solution

$$(3) \qquad w_t = -\frac{a}{b} + \left(w_0 + \frac{a}{b}\right)e^{-bt}$$

where w_t is the soil moisture at time t and w_0 is the soil moisture at $t = 0$.

It is evident that (3) yields $w = w_0$ for $t = 0$ and lim $w = -a/b$ for $t \rightarrow \infty$; i.e., $-a/b$ represents the PWP.

It was further assumed that the parameters of this function vary according to the soil layer and the month. Accordingly the following functions were specified:

$$(4) \qquad ET_{ij} = -dw_{ij}/dt = a_{ij} + b_{ij}w_{ij} \qquad a_{ij} < 0, \quad b_{ij} > 0$$

where j is the index of month, $j = 1, 2, ..., 5$, $j = 1$ denoting December, $j = 2$ denoting January, and so forth; and i is the index of soil layer, $i = 1, 2, ..., 5$. Layers considered had a depth of 30 centimeter (cm) each.

The 'month' indirectly represents the atmospheric evaporative conditions. Note that an alternative formulation of the model was attempted in which ET was expressed as a function of soil moisture w and class A pan evapotranspiration E_0. Although the alternative formulation seemed more attractive conceptually, its predictive power was slightly lower than that of the model applied here.

The parameters a_{ij} and b_{ij} were estimated by a computer search technique to achieve a good fit between the computed moisture values and measured values when they were available. The search was intended to minimize (or rather to obtain low values of) the mean of the absolute values of the relative deviations D between the measured and computed soil moisture values, defined as

$$(5) \qquad D = \left(\frac{1}{\sum_{i,j} K_{ij}} \right) \sum_{k,i,j} \left| D_{kij} \right|$$

with

$$(6) \quad \left| D_{kij} \right| = \left(\left| w_{kij} - \hat{w}_{kij} \right| / w_{kij} \right) \times 100$$

where

k = index of moisture measurement in layer i, month j, $k = 1, 2, \ldots, K_{ij}$; $i = 1, 2, \ldots, I$; $j = 1, 2, \ldots, J$;

K_{ij} = number of moisture measurements in layer i month j;

w_{kij} = measure soil moisture;

$\hat{w}_{kij}$ = computed soil moisture;

$\left| D_{kij} \right|$ = relative absolute value of deviation between the measured and computed moisture values.

(The mean absolute value of relative deviations was chosen as a criterion rather than the conventional standard deviation criterion,

$$\left(\frac{1}{\sum_{i,j} K_{ij}} \sum_{k,i,j} \left(w_{kij} - \hat{w}_{kij} \right)^2 \right)^{1/2}$$

to reduce the relative weight of some extraordinary and out of range measurements that could be the result of error in moisture measurement. Note that the accuracy of soil moisture measurements is generally poor.)

The search was performed in several stages. First, the coefficients of the first layer for all months were considered and then those of the second layer for all months were considered, and so forth. This stage-by-stage procedure seemed more practical than a simultaneous search for all coefficients, which *a priori* should provide better estimates but is more tedious from the computational point of view.

Experimental data for the FA wheat variety in the year 1967/1968 were used for the estimation. There were six treatments with four replications per treatment in this experiment; all together there were 24 plots. Approximately seven moisture measurements per plot during the season were available, including the starting point. The estimates of the parameters a_{ij} and b_{ij} are shown in Table 4.2. Mean values of the absolute relative deviations are shown in Table 4.2. The overall mean relative deviation for all months and plots was 6.2 percent for the three upper layers and 9.3 percent for the five layers. In this context note that when conventional methods are used the coefficient of variation in measurements of soil moisture under experimental conditions; in the order of 10-15 percent. As an illustration, the estimated soil moisture curve for the FA wheat variety T_6 treatment in 1968/1969, layer 1 (0-30 cm), is shown in Figure 4.2.

The parameters a_{ij} and b_{ij} for the FA wheat grown in 1967/1968 were used to reproduce the soil moisture curves for other varieties and years. The seasonal means of the absolute values of the mean relative deviations between the observed and computed moisture values for three and five layers are shown in Table 4.2.

On the basis of the results presented in Table 4.2 it seems that the model for estimating soil moisture is fairly satisfactory with respect to wheat grown in Gilat, having an emergence date in early December and can be used to predict the variations of soil moisture for wheat under similar conditions.

3. RESPONSE FUNCTIONS OF WHEAT YIELD TO SOIL MOISTURE

The soil moisture simulation model described in the previous section was applied for estimating variations in soil moisture in all experimental plots and in all years for the intervals of time between any two consecutive measurements of soil moisture. Corresponding to any date for which a soil measurement was available, computations (of the predicted moisture) were discontinued and started again from the point of the given measurement. Daily computed soil moisture data were used in the estimation of wheat (variety N46) response functions to moisture.

Several soil moisture indices defined on the basis of the daily moisture estimates were tried as potential explanatory variables in the wheat moisture response function. These indices included mean seasonal soil moisture stress in the root zone, mean seasonal soil moisture, mean soil moisture content prior to irrigation, and number of days with soil moisture below a critical value. Construction of separate indices for the different soil layers was attempted. Also the entire growth season of wheat was divided into four growth stages, and separate indices for each stage were defined and tested as explanatory variables. In all 4 years of the experiment the wheat was sown at the end of November. Under the climatic conditions prevailing at the Gilat Experiment Station no rain falls in the summer. While the rainfall season normally begins between mid-November and mid-December, the date of the first rain in the season fluctuates considerably. Germination and emergence of wheat is affected either by rainfall or by irrigation.

Table 4.2. Empirical estimates of the parameters a_{ij} and b_{ij} for the FA wheat variety grown in 1967/1968 using the function $ET_{ij} = a_{ij} + b_{ij}w_{ij}$

	Soil Layer Depth, cm									
	0-30		30-60		60-90		90-120		120-150	
Month	a_{ij}	b_{ij}	a_{ij}	b_{ij}	a_{ij}	b_{ij}	a_{ij}	b_{ij}	a_{ij}	b_{ij}
December	-0.2040	0.0408	-0.0307	0.0041	-0.0307	0.0041	-0.0007	0.0001	-0.0007	0.0001
January	-0.2110	0.0422	-0.1057	0.0141	-0.0007	0.0001	-0.0007	0.0001	-0.0007	0.0001
February	-0.2560	0.0493	-0.1305	0.0174	-0.0652	0.0087	-0.0087	0.0166	-0.0007	0.0001
March	-0.2970	0.0594	-0.4117	0.0549	-0.3202	0.0427	-0.3427	0.0452	-0.0007	0.0001
April	-0.4000	0.0800	-0.1440	0.0192	-0.2400	0.0320	-0.2160	0.0368	-0.0480	0.0064

According to the date of the first rain or the first irrigation, the dates of germination and emergence of the wheat in the experimental plots varied according to year and treatment. The earliest date of emergence for all plots in all years was December 1, and the latest was February 6. The date of emergence was found to be a factor that significantly affected the yield level.

Table 4.3. Mean values of absolute relative deviations for the FA wheat variety in 1967/1968

	Soil Layer Depth, cm					Mean,	Mean,
Month	0-30	30-60	60-90	90-120	120-150	Three Layers	Five Layers
December	3.2	4.3	2.8	7.5	1.6	3.4[b]	3.9[c]
January	5.4	4.6	10.1	12.1	9.9	6.7	8.4
February	4.9	2.9	8.1	18.0	17.2	5.3	10.2
March	8.3	8.1	6.6	15.9	19.0	7.7	11.6
April	5.9	7.8	8.4	14.0	21.1	7.4[b]	11.4[c]
Mean, all months	5.8[a]	5.6	7.3	13.8	14.0[a]	6.2[d]	9.3[d]

[a]Defined as $\left(1/\Sigma_j K_{ij}\right)\ \Sigma_{k,i}\left|D_{kij}\right|$

[b]Defined as $\left(1/\Sigma_i^{\,I} K_{ij}\right)\ \Sigma_{k,i}\left|D_{kij}\right|,\ I = 3$

[c]Defined as above with $I = 5$

[d]D with $I = 3$ or 5, respectively

After extensive examination of various expressions of the response function, the following general functional forms were found to be the best applicable (in terms of the statistical significance of the estimates and the underlying theoretical considerations):

(7) $$Y_i = f_1\left(w_{12}, T, \alpha_i\right)$$

or

(8) $$Y_i = f_2\left(w_{0-9}, w_{9-12},\ G,\ \alpha_i\right)$$

where

Y_i = wheat yield per land unit area in year i, kg/dunam;

w_{12} = number of days during the growth season with soil moisture above 12 percent (about 45 percent of available soil moisture) in the root zone (on the basis of the soil moisture curves it was assumed that the depth of the root zone varied during the growth season as follows: the first 10 days after germination, 30 cm; from the eleventh day after germination to the end of February, 60 cm; from March 1 to April 30, 90 cm);

$w_{0\text{-}9}$, $w_{9\text{-}12}$ = number of days during the growth season with soil moisture below 9 percent and in the range 9-12 percent, respectively;

T = the period in days from the date on which sufficient moisture for germination was available to the latest germination date in the experiment (February 6), an indirect definition of the germination date;

G = the period in days from the earliest germination date in the experiment (December 1) to the actual date for which sufficient moisture for germination was available;

α_i = the effect of the *i*th year, e.g., i = 66, referring to the 1965/1966 season.

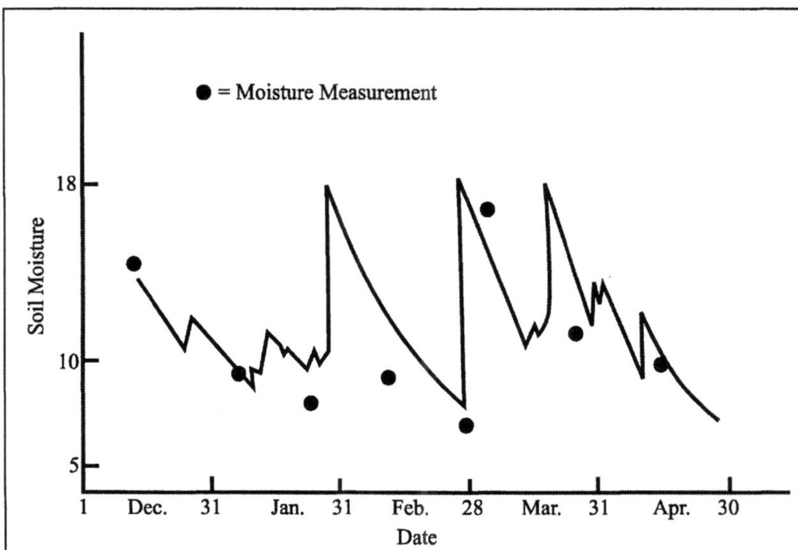

Figure 4.2. Estimated soil moisture variation curve and observed moisture points for wheat variety FA, treatment T_6, in year 1968/1969, layer 1 (0-30 cm)

The 12 percent soil moisture as a critical level was selected on the basis of the slope of the soil moisture tension curve, which becomes flatter at about this point, and because this soil moisture approximately corresponds to a soil suction of 0.75 atmospheres. However, no significant difference between the 12 percent critical level versus, say, 11.0 or 13 percent can be claimed. Obviously the number of days with soil moisture above 11.0 or 13.0 percent is highly correlated with the number of days with soil moisture above 12 percent. Critical moisture values of 11.5 and 13.5 percent were also tried but were found to yield statistical results (in terms of the sum of squares of the unexplained residuals from the estimated response surface) inferior to those at the 12 percent level.

Table 4.4. Seasonal means of the absolute value of the relative deviations D for all varieties and years

		Mean Relative deviation D, %	
Variety	Year	Three Layers 0-90 cm	Five Layers 0-150 cm
FA	1965/1966	13.5	12.1
46	1965/1966	14.2	11.9
FA	1966/1967	13.6	13.1
46	1966/1967	12.9	13.3
FA[a]	1967/1968	6.0	9.5
FA[b]	1967/1968	6.2	9.3
46	1967/1968	6.3	8.8
FA[a]	1968/1969	12.1	13.1
46	1968/1969	11.8	13.5

[a] FA wheat variety chemically treated by chloroethyl trimethyl ammonium chloride
[b] The experiment used for the estimation of the parameters a_{ij} and b_{ij}

Three alternative formulations of the response function along with their empirical estimates are presented below.

3.1. Cobb-Douglass Function

This function is $\ln y_i = a + b \ln w_{12} + c \ln T + \alpha_i$, with the empirical estimate (the asterisks indicating significance at the 1 percent probability level):

$$\text{(9)} \qquad \ln Y_i = 1.775* + 0.591 \ln w_{12} + 0.391* \ln T + \alpha_i$$

and $\alpha_{66} = \alpha_{67} = \alpha_{68} = 0$, $\alpha_{69} = -0.263*$, and $R^2 = 0.97$. The estimate (9) was derived in two stages by using standard regression techniques. At first a model with a term-of-year effect α_i for each year of the observations (i = 66, 67, 68, 69) was formulated and estimated. The estimates of α_{66}, α_{67} and α_{68} were found to be insignificant at an acceptable probability level. On the basis of these results a second model was formulated with a term-of-year effect to be estimated for 1969 only, whereas α_{66}, α_{67} and α_{68} were set to equal 0. The estimate (9) was obtained according to this formulation. Note that in 1969 the experimental field was affected by rust disease, which reduced the yields. Six percent of the variability in $\ln Y$ was due to α_{69}.

3.2. Exponential Function

This function is $Y_i = b_0 b_1^{w_{0-9}} b_2^{w_{9-12}} b_3^{G} \alpha'_i$ with the empirical estimate in logarithmic transformation

(10) $$\ln Y_i = 6.73529 - 0.1500 * w_{0-9} - 0.00793 * w_{9-12} - 0.01589G * + \alpha'_i$$
$$\alpha'_{69} = -0.25346 * \quad \alpha'_{66} = \alpha'_{67} = \alpha'_{68} = 0, \text{ and } R^2 = 0.88$$

The antilogarithmic transformation of (10) is

(11) $$Y_i = 842(0.985)^{w_{0-9}}(0.992)^{w_{9-12}}(0.984)^{G} \cdot \exp(\alpha'_i)$$

The estimate of (10) was derived by regression techniques in two stages, similar to the way (9) was estimated. In the first stage it was found that α_{66}, α_{67} and α_{68} do not differ significantly from 0, and accordingly a second model with year effect only for 1969 was formulated and estimated.

3.3. Mitscherlich Function

This function is

(12) $$Y_i = A[1 - B_1 \cdot \exp(-k_1 w_{12})] \cdot [1 - B_2 \exp(k_2 T)] \cdot \exp(\alpha''_i)$$

with the empirical estimates being

$$Y_i = 700[1 - 0.945 \cdot \exp(-0.015 w_{12})][1 - 0.750 \cdot \exp(-0.0287T)] \cdot \exp(\alpha''_i)$$
$$\alpha''_{69} = -0.415 \quad \overline{R}^2 = 0.91$$

and α''_{66}, α''_{67} and α''_{68} were assumed to be 0 on the basis of previous estimates.

The parameters of the Mitseherlich function were estimated by using the 'steepest ascent' (computer) search technique intended to minimize the sum of squares of the unexplained residuals; $\overline{R}^2$ was computed as

$$\overline{R}^2 = 1 - \left[\Sigma(Y_{ij} - \hat{Y}_{ij})^2 / \Sigma(Y_{ij} - \overline{Y})^2\right]$$

with the conventional notation. No tests of significance of the parameters for the Mitscherlich function were performed, owing to the lengthy computations involved and the information gained from the test of significance of the coefficients of the Cobb-Douglass and exponential functions.

All three estimates (9), (10), and (12) yielded satisfactory results from the statistical point of view. With regard to the conceptual assumptions underlying the three formulations (assumed to be known to the reader and not repeated here) the Cobb-Douglass function was considered inferior. The empirical estimate of the exponential function (10) gave 842 kg/dunam as the asymptotic yield obtainable under optimal conditions, with $w_{0-9} = w_{9-12} = G = \alpha_i = 0$, which seemed too high to wheat specialists, whereas the Miltscherlich function (12) gave 700 kg/dunam as the asymptotic yield obtainable under optimal conditions, which seemed quite reasonable. On the basis of these considerations the Mitscherlich function was chosen as the 'best estimate' and as the basis for the analysis of optimal irrigation policy. The variation of soil moisture in the experimental plots was simulated, and the corresponding yields were estimated by using the alternative estimates (9) and (12). The estimated yield values $\hat{\hat{Y}}_{ij}$ (i = 66, ..., 69; j = 1, 2, ..., 6) were compared with the actual yield values Y_{ij}, and the mean of relative deviations defined as

$$\left\{(1/IJ)\Sigma_{ij}\left[\left(Y_{ij} - \hat{\hat{Y}}_{ij}\right)/Y_{ij}\right]\right\} \times 100$$

was computed. It was found to be 7.8 percent by using the Mitscherlich function (12) and 12.3 percent by using the Cobb-Douglass function (9). Owing to time limitations, such analysis was not performed with respect to the exponential function.

4. OPTIMAL IRRIGATION POLICY

Two major questions are apparently of interest with respect to irrigation policy. What is the optimal irrigation policy under conditions similar to those of the site of the experiment? To what extent can irrigation decisions be improved by detailed information on soil moisture, such as that provided by the soil moisture estimation model?

To obtain answers to these questions, the soil moisture estimation model and the estimate of the Mitscherlich response functions with $\alpha_i'' = 0$ were applied in the following analysis, in which the rainfall record at the site of the experiment for a period of 16 years was taken into account. This record was taken as representing the random rainfall distribution in the area.

At first, two approaches to irrigation decisions were tested in a simulation analysis: (1) irrigating on the basis of a predetermined time schedule, the quantities of water applied being equal to the moisture depletion in the root zone at the time of irrigation and (2) irrigating at the date on which the soil moisture is depleted to a predetermined critical level.

Several irrigating policies derived from the approaches above have been tested by simulating their consequences in relation to the 16-year rainfall record. The corresponding expected values for water use, wheat yield, and net revenue per unit area of land are shown in Table 4.5 for two levels of water cost, i.e., IL 0.10/m^3 and IL 0.20/m^3 (1 IL = Israeli Lira is about US$0.24). The water cost included the cost of application and allowances were made for conventional delivery losses to the field. All policies presented in Table 4.5 assume knowledge of soil moisture depletion at the time of irrigation.

Table 4.5. Expected values of water use, yield of wheat, and net return per unit area of land under selected irrigation policies based on computed soil moisture depletion values at time of irrigation

Irrigation Policy	Critical Level, %	Dates of Irrigation	Water Use m^3/ dunam	Yield kg/ dunam	Net Return,[a] IL/dunam Water Cost 0.10 IL/m^3	Water Cost 0.20 IL/m^3
1. No irrigation			0	225	61 (50)	61 (50)
		Irrigation to Predetermined Time Schedule				
2. One		Dec 1	150	437	125 (20)	110 (25)
application		Jan 1	100	342	99 (26)	89 (31)
3.						
4. Two		Dec 1, Jan 1	200	463	128 (12)	108 (24)
applications		Dec 1, Feb 1	200	480	134 (10)	114 (25)
5.		Dec 1, Mar 1	230	474	129 (15)	106 (29)
6.						
7. Three		Dec 1, Jan 1, Feb 15	255	493	132 (7)	107 (28)
applications		Dec 1, Jan 1, Mar 1	265	494	132 (10)	105 (31)
8.		Dec 1, Jan 15, Mar 1	270	503	134 (8)	107 (30)
9.		Dec 1, Feb 1, Mar 15	310	514	134 (11)	103 (36)
10.						
		Irrigation according to Measured Moisture Level				
11.	9		150	440	125 (20)	110 (28)
12.	10		190	460	128 (14)	109 (26)
13.	11		210	482	133 (10)	112 (26)
14.	12		220	500	138 (7)	116 (26)

[a] The numbers in parentheses are estimates of population standard deviations. To obtain standard deviations of the means, divide these numbers by 4 ($n^{1/2}$; $n = 16$).

Scrutiny of the data presented in Table 4.5 suggests that the best policy (policy 14) is to irrigate whenever the moisture level in the root zone is depleted to 12 percent. The second best policy (policy 5) is the one using two predetermined water applications (on December 1 and February 1). However, the difference between the two policies is quite small and statistically nonsignificant (if a normal distribution of the net return values per dunam is assumed as an approximation and the *t*-test is applied to the comparison of the two means). If one adopts the expected variance as a choice criterion and compares both the means and their standard deviations, policy 14 is again slightly better than policy 5; both 14 and 5 are obviously better than policy 1 (no irrigation).

Note that policy 5 uses information on soil moisture depletion before irrigation to determine the quantities of water to be applied. If such information is not available, the difference between policies 14 and 5 will be somewhat larger. For example, overuse (assuming zero benefit) of, say, 50 m^3/dunam will reduce the net, returns by IL 5/dunam or IL 10/dunam according to the two levels of water cost in the system exemplified by policy 5.

Policy 14 is particularly advantageous in years of low rainfall, as is evident from Table 4.6. In such years responsive control-oriented irrigation decisions are apparently justified.

Table 4.6. Comparison of estimated net returns per unit area of land (IL/dunam) under policies 5 and 14 in years of low rainfall

		Policy 5 Net Return, IL/dunam		Policy 14 Net Return, IL/dunam	
Rainfall, mm/dunam	Year	Water Cost 0.10 IL/m^3	Water Cost 0.20 IL/m^3	Water Cost 0.10 IL/m^3	Water Cost 0.20 IL/m^3
42	1963	113	88	130	100
73	1960	118	94	127	97
133	1962	135	112	135	112
34	1966	126	103	141	119
154	1854	133	110	128	98
168	1955	126	104	128	98
180	1969	130	110	136	113

5. CONCLUSION

A procedure for estimating soil moisture under wheat production and estimates of response functions of wheat yield to soil moisture have been presented and a simulation analysis of irrigation policies has been discussed. The analysis indicates preferred irrigation policies. The benefit of the soil moisture estimation model in making irrigation decisions is evident primarily with respect to years with low rainfall. Since application of the moisture estimation model is very simple and almost costless, once the relevant parameters are estimated, farmers and irrigation specialists may benefit from using such a model.

Although the approach presented is general, note that the parameters estimated are only valid locally. Further study is needed to evaluate similar parameters for different conditions and other crops.

The analysis ignores the restrictions faced by farmers in adopting the best policy under real-life conditions (such as shortage of water supply, lack of equipment and manpower, and so forth). However, the approach presented can be extended to the analysis of optimal irrigation decisions within a real-life farm system.

REFERENCES

Baier, W. and G. W. Robertson. Estimating yield components of wheat from calculated soil moisture. *Canadian Journal of Plant Science*, 47:617-629. 1967.

Blaney, H. F. and W. D. Criddle. Determining water requirements in irrigated areas from climatological and irrigation data. SCS-TP 96. United States Department of Agriculture, Soil Conservation Service. Washington, DC. 1950.

Burt, O. R. and M. S. Stauber. Economic analysis of irrigation in subhumid climate. *American Journal of Agricultural Economics*, 53:33-46. 1971.

De Lucia, R. J., Operating policies for irrigation systems under stochastic regimes (PhD. Thesis). Harvard University. Cambridge, MA. 1969.

Dorfman, R. Response of agricultural yields to water in the former Punjab. Report on land and water development in the Indus Plains, Appendix A-5. White House Panel on Water Logging and Salinity in West Pakistan. Washington, DC. 1963.

Dudley, N. J., D. T. Howell and W. F. Musgrave. Optimal interseasonal irrigation water allocation. *Water Resources Research*, 7:770-788. 1971a.

Dudley, N. J., D. T. Howell and W. F. Musgrave. Irrigation planning, 2. Choosing optimal acreages within a season. *Water Resources Research*, 7:1051-1063. 1971b.

Flinn, J. C. and W. F. Musgrave. Development and analysis of input-output relations for irrigation water. *Australian Journal of Agricultural Economics*, 11:1-19. 1967.

Hall, W. A. and W. S. Butcher. Optimal timing of irrigation. *Journal of Irrigation and Drainage Division American Society of Civil Engineers*, 94(IR2):267-275. 1968.

Jensen, M. E. Empirical methods of estimating or predicting evapotranspiration using radiation. "Proceedings of Water Resources Management Conference." American Society of Agricultural Engineers. 1967.

Moore, C. V. A general analytical framework of estimating the production function for crops using irrigation water. *Journal of Farm Economics*, 43:876-888. 1961.

Penman, H. L. The dependence of transpiration on weather and soil conditions. *Journal of Soil Science*, 1:74-89. 1949.

Shaw, R. H. Prediction of soil moisture under meadow. *Agronomy Journal*, 56:320-324. 1964.

Thornthwaite, C. W. and J. R. Mather. The water budget and its use in irrigation in: "Water." United States Department of Agriculture. 346-357. 1955.

Weisbrod, M., G. Strateener, D. Yaron, D. Shimshi and E. Bresler. Simulation model of water variation in soil. Research Paper, 1201. Center for Agricultural Economic Research. Rehovot, Israel. 1971. (Hebrew)

Yaron, D. Estimation and use of water production functions in crops. *Journal of Irrigation and Drainage Division American Society of Civil Engineers*, 97(IR2):291-303, 1971.

Yaron, D., H. Bielorai, U. Wachs and J. Putter. Economic analysis of input-output relations in irrigation. "Transactions of the Fifth Congress in Irrigation and Drainage." International Commission on Irrigation and Drainage. New Delhi, India. 3:13-16. 1963.

5

A MODEL FOR THE ECONOMIC EVALUATION OF WATER QUALITY IN IRRIGATION*

Dan Yaron
The Hebrew University of Jerusalem, Rehovot, Israel

Eshel Bresler
The Volcani Institute of Agriculture Research, Bet Dagan, Israel

1. INTRODUCTION

Many investigations have shown that plants respond to the concentration of salts in the soil solution of the root zone (Allison, 1964; Bernstein, 1964; Bernstein and Hayward, 1958; Black, 1968; Grillot, 1954; Hayward, 1954; Hayward and Bernstein, 1958; Magistad, 1945; Hayward and Wadleigh, 1949; Magistad et al., 1943). The response may arise from a high total concentration of salts (osmotic effects), a relatively high concentration of a specific ion, or a combination of the two. The specific effect of an ion may be one of direct toxicity or of nutrition. Depending on the nature of the particular plant and given conditions of growth, one of the above two effects is dominant. Another effect is that of a gradual perennial accumulation of salts in the soil, which may lead to soil deterioration in the long run, even though the effects with respect to the concurrent crops are negligible.

Most fruit trees, particularly citrus, are specifically sensitive to the concentration of chloride ion in the soil solution (Bernstein, 1964, 1965). Consequently, in Israel—with citrus as a major product—the chloride concentration is considered to be the main factor determining the suitability of water for irrigation. The chloride concentration is referred to as the measure of salinity in this study, and the terms salinity and chloride concentration will be used in-

*Permission to publish this chapter was granted by the Australian Agricultural and Resource Economics Society. The chapter was originally published under a similar title in *The Australian Journal of Agricultural Economics,* 14 (June):53-62, 1970.

terchangeably throughout this chapter. At the same time the approach of the model is adequate in reference to other measures of salinity, such as total salt concentration and electrical conductivity of the soil solution.

Information on the processes of accumulation and leaching of salts in irrigated soils and of the response of plants is a prerequisite for the management of an orderly irrigation regime, when saline water is used. It is generally accepted by soil scientists that the salt concentration of the soil solution, rather than that of the irrigation water, is the ultimate factor, which affects crop yields (Allison, 1964; Bernstein 1964; Bernstein and Hayward, 1958; Bresler, 1967; Hayward, 1958; Hayward and Bernstein, 1958; Hayward and Wadleigh, 1949). By increasing the quantity of water applied per land unit, a portion of the salt in the soil solution can be leached below the root zone, and consequently, irrigation water with a higher salinity rate may be used. Here the question arises what is the 'best' combination of water quantity and quality in irrigation under particular field conditions?

2. FRAMEWORK OF THE ANALYSIS AND OBJECTIVES

An economic evaluation of the 'best' combination of water quantity and quality implies knowledge of the production function which relates the crop yield per land unit to varying levels of the above two factors. In a most general form such a function may be written as:

(1) $$Y = f(Q, S \mid K)$$

where

Y = crop yield per unit area of land;
Q = quantity of water of standard quality applied per unit area of land;
S = index of salt concentration in the soil solution, during the growing period; with
K representing all other factors, assumed to be constant.

At this stage we confine ourselves to a rather vague definition of S; a precise definition will be presented in the following sections of the chapter.

The salt concentration of the soil solution is itself a function of other variables, which, in a general way, can be expressed as:

(2) $$S = g(S_0, Q, C \mid K)$$

where

S_0 = index of initial salinity conditions;

C = salt concentration of the irrigation water applied.

Substituting (2) into (1) we obtain[1]

(3) $$Y = h(S_0, Q, C \mid K).$$

The model assumes the existence of a variety of water resources with different levels of salinity and cost of supply. Such a situation prevails in Israel, where water suppliers (mostly regional and national, and some private suppliers) face the problem of increasing cost of water supply with decreasing salt content. The objective of this chapter is to present an approach for the derivation of the optimal combination of quality and quantity of water under given conditions of climate, soil, land use, and the relative cost of water quantity and quality, subject to restrictions on salt concentration in the soil solution. The ultimate relationship of interest, in this context, is function (3). Such functions have been the subject of much research, but are not included in this study. Here the focus is restricted to developing an approach for the estimation of functions like type (2).[2] More specifically, a model is designed within the framework of which:

a) S_0 and K of (2) are considered as given;
b) S—the index of salt concentration in the soil solution during the irrigation period is restricted not to exceed a predetermined critical level;
c) efficient combinations of Q and C which comply with the above conditions are estimated.

The rationale for estimating functions of type (2) originates in previously established research (Bernstein, 1964, 1965; Bierhuizen, 1969), where significant yield response to salinity was observed only above a critical threshold concentration. Below the threshold concentration response is negligible. Accordingly, a partial analysis based on functions of type (2) may provide practical guidance for irrigation policy.

Following the discussion of the conceptual framework of the study we present in the next section the essence of the economic framework of the analysis; in the fourth section the physical relationships underlying the model are presented and in the fifth section the approach to the evaluation of the optimal combination of water quantity and quality is demonstrated. An attempt to evaluate the results and to point out possible extensions of the study concludes the presentation.

3. THE ECONOMIC FRAMEWORK

In this section we introduce some elementary concepts needed for the determination of the economically optimal combination of water quality and quantity in irrigation. Let our starting point be function (2) of the previous section with constant S_0 and K:

(4) $$S = g(Q, C \mid S_0, K)$$

Let us consider those combinations of Q and C which yield the same soil salinity index at the end of the irrigation period, denoted, say, by S_1 and draw them on a graph in two dimensions, the axes being Q—the quantity of water applied in irrigation, and C—the salinity of water, with the degree of salinity increasing along the ordinate towards the origin (Figure 5.1). Denote the curve, which represents the above combinations of Q and C by S_1S_1. Available information regarding the processes of salt accumulation and leaching suggests that over a certain range of this curve (in which salt is leached by excessive amounts of water) Q and C vary in the same direction, namely the same index of soil solution salinity may be achieved by simultaneously increasing (or decreasing) both the quantity of water and its salt concentration. In the following we shall refer to S_1S_1 as the 'iso-soil-salinity' curve at the S_1 level. Note that other iso-soil-salinity curves will correspond to other levels of S.

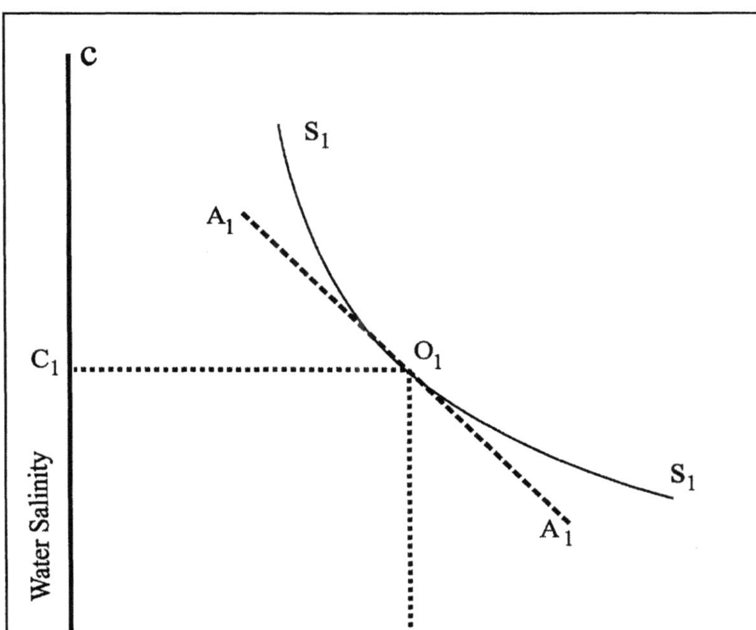

Figure 5.1. *A hypothetical iso-soil-salinity curve and determination of the optimal quantity-salinity concentration combination*

Our problem is to find the least cost combination of water quantity and salinity, which provides for a given soil salinity index S_1, at the end of the irrigation period. It is assumed that (a) the water supplier can control the level of water salinity by mixing water from different sources; (b) cost of water supply increases with the decrease in its salt content.

Denote:

M = salt concentration of irrigation water of standard salinity in parts per million (ppm);

p_q = cost per water unit of standard salinity ($/m^3);

p_c = cost of deviating the salt concentration of one water unit by one ppm from the standard salinity ($/m^3/ppm).

M is referred to as a base reference concentration; for $C > M$, p_q is adjusted downwards; for $C < M$, it is adjusted upwards. Such an adjustment can be thought of as a premium received or paid, respectively, by the water user, according to the quality of water. In the following, Q is measured in m^3 and S_1 in ppm.

To derive the optimal quantity-quality combination of a system as in Figure 5.1 we form the following Lagrangian expression:

$$L = Q \cdot P_q + (M - C) \cdot p_c \cdot Q - \lambda\left[g\left(Q,\ C\right) - S_1\right]. \tag{5}$$

Taking partial derivatives with respect to Q, C, λ and equating to zero we get

$$\begin{aligned} \partial L/\partial Q &= p_q + (M - C) \cdot p_c \quad - \lambda \cdot \partial g/\partial Q = 0 \\ \partial L/\partial C &= -p_c \cdot Q \quad - \lambda \cdot \partial g/\partial C = 0 \\ \partial L/\partial \lambda &= -g\left(Q,\ C\right) \quad + S_1 = 0 \end{aligned} \tag{6}$$

which simplifies to

$$\begin{aligned} dC/dQ &= \left[p_q + (M - C) \cdot p_c\right] / p_c \cdot Q \\ g\left(Q,\ C\right) &= S_1 \end{aligned} \tag{7}$$

As shown in (7) the necessary condition for the least cost quantity-quality combination is that the marginal rate of substitution of water salinity for quantity be equal to the ratio of the cost per water unit adjusted for quality to the cost of varying the concentration of the given quantity of water by one ppm.[3] The least cost quantity-quality combination is represented graphically in Figure 5.1, with S_1S_1 representing the iso-soil-salinity curve at the S_1 level, the line segment A_1A_1 representing the cost of quantity and quality ratio and the point O_1—the optimal combination.[4] Clearly, if

$$(8) \qquad dC/dQ > \left[p_q + (M - C)\cdot p_c\right] / p_c \cdot Q$$

at any point on the S_1S_1 curve, it pays to simultaneously increase water quantity and salinity. A move in the opposite direction is justified if the inequality sign in (8) is reversed. Note also that for any level of 'iso-soil-salinity' conditions an optimal *Q-C* combination can be derived similarly.

The determination of the desired isosalinity level depends on the knowledge of the empirical functions (3) previously discussed. If these functions are not known, the isosalinity level may be determined at the level of the threshold concentration. Evaluation of the threshold concentration regarding citrus is provided by Bernstein (1965): "For example, the chloride concentration in the saturation extract of the soil, with sweet orange or citrange roots, should not exceed 10 meq/l or 350 ppm. Trees on sour orange, rough lemon or tangelo roots, however, will not be damaged if the chloride is less than 15 meq/l in the saturation extract. On Rangpur lime and Cleopatra mandarin roots, trees will tolerate up to 25 meq/l chloride."

4. MODEL FOR TRACING SALT DISTRIBUTION IN THE SOIL

In this section we present a model for tracing salt distribution in the soil profile under varying field irrigation conditions.[5] The model is essentially a reformulation of the law of mass conservation, and states that the amount of salt added in irrigation to any of the soil layers, less the amount leached and the amount absorbed by the plants, is equal to the net increment (positive or negative) of salt in this layer. Assuming that the amount of salt absorbed by the plant is negligible, and that the salt concentration of the water, which is leached through any soil layer is equal to the arithmetic mean of the concentrations before and after irrigation, the basic relationship of the model for one layer and a single water application can be written as

$$(9) \qquad QC - (Q - E)(X_0 + X_1)/2 = (X_1 - X_0)/\theta D$$

where

Q = depth of irrigation water applied (mm);
C = chloride concentration in the irrigation water (meq/l)[6];
E = water consumption (evapotranspiration) from the relevant soil layer (mm);
θ = water content (volume fraction) at time of extraction for chloride analysis (cm^3/cm^3);
X_0 = initial chloride concentration in the soil solution at water content θ (meq/1);
X_1 = chloride concentration in the soil solution, at water content θ, after irrigation (meq/l);
D = depth of the relevant soil layer (mm).

Note that QC represents the amount added, $(Q - E)(X_0 + X_1)/2$ the amount drained, and $(X_0 - X_1)$¬D the net increment of salt in the relevant soil layer.

Equation (9) can be extended so as to enable tracing of salt distribution over any number of soil layers, in response to any number of successive irrigations. For the rth soil layer ($r = 1, 2, \ldots, m$) and the jth successive irrigation applied ($j = 1, 2, \ldots, n$), we obtain

$$(10) \quad Q_jC_j - \left(Q_j - \sum_{i=1}^{r} E_{ij}\right)\left(X_{r,j-1} + X_{rj}\right)/2 = \sum_{i=1}^{r}\left(X_{ij} - X_{i,j-1}\right) W_i$$

where $W_i = \theta_i D_i$.

Thus, for m soil layers and n irrigations, a system of $m \times n$ linear equations is obtained, solvable for the X_{ij} by routine methods.

The validity of this salt tracing model and its predictive power was tested using data obtained from three citrus irrigation experiments carried out at three locations in Israel—Gevat, Shefayim, and Gilat. The experimental results, i.e. the actual salt distribution throughout the soil profile during the irrigation process, were compared with the values of X_{ij}*'s* calculated using the model. The comparisons have shown a fairly good fit between the observed and the calculated values. Details of these comparisons are presented in Bresler (1967) and Yaron and Bresler (1966). In view of these results it seems that the above model constitutes a fairly good tool for the approximate estimation and prediction, under field conditions, of accumulation and leaching of chlorides in soils.

5. ESTIMATION OF ISO-SOIL SALINITY CURVES AND DETERMINATION OF THE OPTIMAL WATER QUANTITY AND SALINITY COMBINATIONS

The variables in the system (10) can be classified into three major groups: (a) constants and (directly or indirectly) predetermined variables,[7] such as W_i, X_{i0} and E_{ij}; (b) parameters which are liable to direct man control, such as C_j and Q_j; and (c) dependent variables such as X_{ij}. Of particular interest is the dependence of X_{ij} on the values of the control variables C and Q, which uniquely determine the salinity regime throughout the soil profile during the irrigation process, subject to the values of X_{i0}, W_i and E_{ij}. With these relationships in mind the leaching model can be applied to the analysis of the influence of particular C and Q values on the salt distribution in the soil profile.

In this chapter an application of linear programming and computer simulation to the estimation of iso-soil-salinity curves under field conditions is presented. To this end we formulate the following problem. (a) Let Q_j be predetermined at some specific level. (b) Let the chloride concentration in the soil solution in each soil layer and after each irrigation, be restricted not to exceed 12.5 meq C1/1, equivalent to 444 ppm Cl.[8] (c) Find the maximal level of C which will not violate the restriction (b), with Q_j and other parameters (X_{i0}, W_i, E_{ij}) being given.

In the linear programming formulation the problem takes the following form (find the value of C which maximizes L)

$$(11) \qquad L = C \cdot \sum_{j=1}^{n} Q_j \rightarrow \max$$

subject to:

(i) the set of salt balance equations (10) with given values of E_{ij}, X_{i0}, and W_i ($i = 1, 2, ..., m; j = 1, 2, ..., n$);

(ii) restrictions on the critical levels of chloride concentration in the soil solution:

$$X_{ij} \leq 12.5 \qquad i = 1, 2, \ldots, m$$
$$j = 1, 2, \ldots, n.$$

Note that the problem consists of a linear function (11) to be maximized subject to a set of linear equalities and inequalities (i) and (ii). The solution provides for the maximal permissible value of C and the resulting values of X_{ij}. By successively varying the predetermined values of Q_j and solving for C accordingly, a set of combinations of ΣQ_j with the maximal permissible water salinity C is achieved, which maintain the iso-soil-salinity conditions,[9] specified by $X_{ij} \leq 12.5$.

Another approach to the solution of the same problem is based on application of a computer simulation code designed for solving the leaching model

(10) for X_{ij} subject to variations in the control parameters C and Q_j. The maximal permissible value of C is found by computerized trial and error.[10]

For sake of illustration we present the essentials of an application of this model to the analysis of irrigation of citrus groves at Gevat (Israel). Irrigation experiments performed there provided the general framework for the analysis. Five layers of 30 cm each, and six irrigations were considered. The initial salinity values (X_{i0}) were assumed to be 7.0, 8.0, 8.5, 9.0, and 9.5 meq Cl/l, equivalent to 249, 284, 302, 320, and 337 ppm C1, in the five layers respectively.[11] The totality of water applied in the six irrigations was 663 mm and the critical salinity in the soil solution was set not to exceed 12.5 meq Cl/l; equivalent to 444 ppm Cl.[12] The maximum permissible chloride concentration in the irrigation water was found to be 202 ppm Cl.

Next, variants of the problem, each with different total quantity of water applied, were formulated, the quantities varying from 464 up to 1160 mm (70-175 per cent of the standard quality). The results of this analysis are summarized in Table 5.1, and Figure 5.1. Figure 5.1 presents the iso-soil-salinity curve with respect to the 444 ppm Cl restriction on the salinity of the soil solution. The corresponding (approximate) marginal and average rates of substitution between water salinity and quantity are presented in the last row of Table 5.1. These rates over the entire range of water quantities vary between 0.11 and 0.26 ppm Cl/m^3. The average rate of substitution over the 100-150 percent range of the standard water quantity is 0.21 which means that an addition of one mm in water quantity allows for an increase of about 1/5 ppm Cl in water salinity.

Table 5.1. *Empirical estimates of marginal and average rates of substitution of water salinity for quantity* [a]

Total Quantity of water Applied, mm	464	663	829	995	1160
Percentage of standard	70	100	125	150	175
Maximal permissible water salinity ppm Cl	167	202	245	270	288
Range of variation in water quantity, Q mm	464-663	663-829	829-995	995-1160	663-995
Difference between quantity levels over the range ΔQ mm	199	166	166	166	332
Corresponding difference in salinity level ΔC ppm Cl	35	43	25	18	69
Marginal or average rate of substitution $\frac{\Delta C \text{ ppm Cl/mm}}{\Delta C / \Delta Q \text{ ppm Cl/mm}}$	0.18	0.26	0.15	0.11	0.21

[a] See text for description of the frame of the analysis

In order to derive the optimal combination of water quantity and salinity, the cost of water in terms of quantity and salinity should be known. These, however, have not yet been established in Israel. Therefore, it is possible, thus far, to discuss this problem in general terms only. Using the notation and the frame of analysis of section three, the empirical estimates of the marginal rate of substitution of water salinity for quality (last row of Table 5.1) should be compared with the ratio $[p_q+(M-C)\mathrm{E}p_c]/\, p_c\, \mathrm{E}Q$. The necessary condition for optimal combination is that $dC/dQ = [p_q+(M-C)\mathrm{E}p_c]/p_c\mathrm{E}Q$. If more than one point on the isosalinity curve has this property, the cost attached to each of such points should be compared, and the least cost point be chosen. In the particular example of Figure 5.2, the isosalinity curve is close to linear and a corner optimum solution is likely. It should be noted, however, that other isosalinity curves derived in the study for different situations depart definitely from linearity.

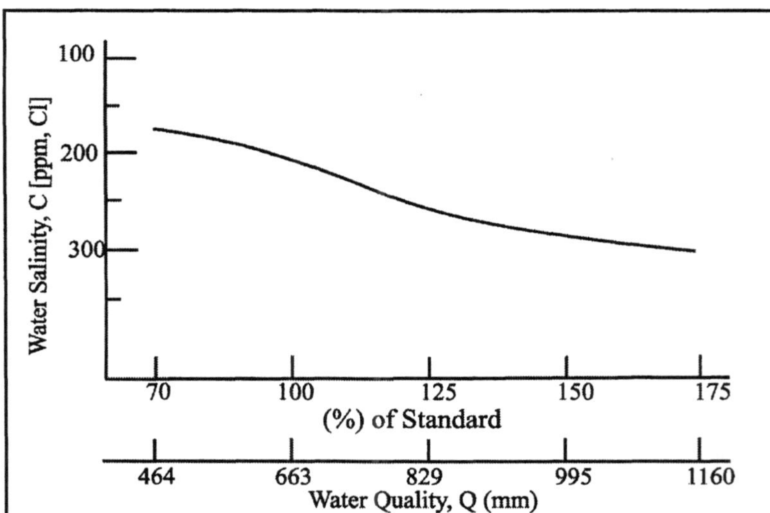

Figure 5.2. *An iso-soil-salinity curve $X_{ij} \leq$ meq Cl/l derived with the aid of the linear programming model*

It is emphasized that the above numerical results are meaningful only with respect to the particular set of conditions referred to in the analysis. They are presented here for sake of illustration only. No attempt is made at this stage to arrive at any generalizations and conclusions regarding the nature of the relationship between water quantity and quality and the use of saline water.

Another question relates to the economically optimal level of the iso-soil-salinity index (specified by the restrictions on X_{ij} in the linear programming model) above the threshold. To provide an answer to this question the relationship between the salt concentration in the soil solution, the quantity of water applied, and the crop yields should be empirically established. A study aimed at this goal is now under way.

6. CONCLUSION

The approach presented in this chapter provides a tool for a partial economic analysis of irrigation practices under conditions of salinity of the irrigation water. The importance of guides that may be derived from such analysis with respect to water supply policy is obvious. The major concern of this presentation is one of exploration of the model and its capabilities; no attempt is made to arrive at any generalizations regarding the nature of the substitution relationship between water quantity and salinity. Considerable further research effort is needed in order to arrive at reasonably reliable empirical estimates of the parameters of interest.

Several extensions and modifications of the model are possible. In particular an extension of the analysis over a series of years, which may account for the year-to-year effects may prove important. It may also be of interest to analyze the possibility of periodical variations in the salt concentration of the irrigation water, the pattern of these variations being derived from the supply conditions.

Finally it should be noted that the present study deals with only one particular aspect of the whole complexity of economic problems involved in the use of saline water in irrigation. An extensive and diversified research effort is needed in order to provide the information necessary for sound policy decisions with respect to water resource development.

REFERENCES

Allison, L. E. Salinity in relation to irrigation. *Advances in Agronomy*, 16:139-178. 1964.

Bernstein, L. Salinity and citrus. *California Citrograph*, 50:273-274. 1965.

Bernstein, L. Salt tolerance of fruit crops. *United States Department of Agriculture Information Bulletin*, 292. 1964.

Bernstein, L. and H. E. Hayward. Physiology of plant tolerance. *Review of Plant Physiology*, 9:25-46. 1958.

Bierhuizen, J. F. Water quality and yield depression. Institute for Land and Water Management. Wageningen, Holland. *Research Technical Bulletin*, 61. 1969.

Black, C. A. "Soil Plant Relationship." John Wiley & Sons. New York, NY. 1968.

Bresler, E. A model for tracing salt distribution in the soil profile and estimating the efficient combination of water quality and quantity under varying field conditions. *Soil Science*, 104:227-233. 1967.

Grillot, G. The biological and agricultural problems presented by plants tolerant of saline or brackish water and the employment of such water for irrigation. *Arid Zone Research*, 4:9-35. 1954.

Hayward, H. E. Plant growth under saline conditions. *Arid Zone Research*, 4:37-71. 1954.

Hayward, H. E. and L. Bernstein. Plant-growth relationship on salt affected soils. *Botanical Review*, 24:584-635. 1958.

Hayward, H. E. and C. H. Wadleigh. Plant growth on saline and alkali soil. *Advances in Agronomy*, 1:1-38. 1949.

Magistad, O. C. Plant growth relations on saline and alkali soils. *Botanical Review*, 11:181-230. 1945.

Magistad, O. C., A. D. Ayers, C. H. Wadleigh and H. G. Guach. Effect of salt concentration, kind of salt and climate on plant growth in sand cultures. *Plant Physiology*, 18:151-166. 1943.

Yaron, D. and E. Bresler. An economic approach to the analysis of the salinity problem. Research report. Department of Agricultural Economics. The Hebrew University of Jerusalem. Rehovot, Israel. 1966.

---

[1] As only a general frame of the relationships involved is presented here, no reference to indices of time is made. Also the effect of salt accumulation in the soil profile by the end of the irrigation period is temporarily ignored. A detailed model, which allows for a specific treatment of these phenomena, is presented later in the chapter.

[2] Empirical estimation of (3) is a significantly more ambitious goal than that of (2). This is so due to the complexity of (3) relative to (2) and to the relative scarcity of data for estimation of (3).

[3] If the S_1S_1 curve is convex the necessary conditions are also sufficient.

[4] Note that: (a) dC/dQ along a given iso-soil-salinity curve is positive; (b) the cost ratio $[p_q+(M-C)\,p_c]/p_c\,Q$ is not constant and depends on Q and C.

[5] For more details on the model and review of other approaches, see Bresler (1967) or Yaron and Bresler (1966).

[6] 1 meq Cl/l (one millequivalent of chloride per litre = 35·5 mg Cl/l = 35·5 ppm Cl.

[7] It is assumed that over a range of variation in Q_j, in which leaching occurs, the evapotranspiration rate does not change.

[8] This restriction was formulated on the basis of Bernstein's (1965) evaluation regarding threshold salinity of the soil solution with respect to citrus. Other restriction levels in the range of 10-15 meq Cl/l were applied in other analyses.

[9] An alternative specification of these restrictions could be $\sum_{i=1}^{m} X_{ji}/m \leq 12.5$. However, in order to be on the safe side, the more restrictive formulation was chosen.

[10] The present code consumes 1.4 minutes of 7040 IBM computer time to solve 15 sets of a system, each consisting of five soil layers and six irrigations, with 15 alternative values of C.

[11] The actual initial salinity values were, in general, considerably lower (see Yaron and Bressler 1966). The assumed values seemed to comply better with situations, which may develop after several seasons of irrigations under conditions unfavorable for salt leaching.

[12] The shift in units from meq Cl/l to the ppm Cl is parallel to the common practice in Israel; while soil scientists tend to use meq Cl/l economists, policy makers and water suppliers generally refer to ppm.

6

APPLICATION OF DYNAMIC PROGRAMMING IN MARKOV CHAINS TO THE EVALUATION OF WATER QUALITY IN IRRIGATION*

Dan Yaron and Amikam Olian
The Hebrew University of Jerusalem, Rehovot, Israel[1]

1. INTRODUCTION

Considerable experimental work has been carried out on the effect of water quality on crop yields (Bernstein, 1964 and 1965; Bierhuizen, 1969; DeForges, 1970 and Sahlhevet et al., 1969), but relatively few studies on the economic dimension of the subject have been published (MacAvoy, 1969; Pincock, 1969; Timmons and Dougal, 1967 Yaron, 1966; and Yaron and Bresler, 1970). Most of the economic-oriented studies adopt a static framework and neglect the dynamics of the system. Such dynamics play an important role in situations typical to the Mediterranean climate, where a dry summer irrigation season is followed by a rainy winter with stochastic quantities of rainfall. This chapter emphasizes the dynamic aspects of irrigation with saline water over a sequence of years.

In an analysis of the value of water quality in irrigation, a distinction should be made between three ranges of time and three corresponding models: a) A "short run" model refers to relationships confined to a single irrigation season. The model refers to the initial salinity of the soil profile at the beginning of the irrigation season as given for each initial state analyzes the optimal combination between water quantity and quality but does not take into account the effects of accumulation of salt over time. "Short run" irrigation problems are not discussed in detail here (refer to Bresler and Yaron, 1972 and Yaron and Bresler, 1970). b) A "long run" model accounts for the effects of salt accumulation in the soil profile over time. It comprises a

*Permission to publish this chapter was granted by the American Agricultural Economics Association. The chapter was originally published under a similar title in the *American Journal of Agricultural Economics*, 5 (3):467-471, 1973.

succession of short run processes—the initial conditions of which are affected by salt accumulation in previous periods; irrigation decisions over a single season take into account the resulting terminal conditions due to each alternative and the effects in succeeding periods. c) An "extended long run model" takes into account both salt accumulation in the soil profile and its accumulation in the underground water reservoirs.

This chapter is restricted to the "long run" only. It neither explores irrigation decisions in the "short run" nor discusses the "extended long run" effects of salinity accumulation in underground reservoirs.

2. THE LONG RUN MODEL

The basic conception is one of an N stage dynamic system, with a given number (M) of states, a succession of decision options, random events, transformation functions of the system from one state to another, and a profit (or loss) function related to the decision options and the states of the system. The nature of the problem fits well with the dynamic programming model in Markov chains (Howard, 1960).

Two major situations concerning crop types can be distinguished. a) In irrigation of perennial crops, such as citrus, the goal is to determine the optimal combination of water quantity and quality with the crop held constant. Farmers have the option to replace the crop if it is unprofitable over a series of years. b) In irrigation of annual crops, such as field crops and vegetables, the problem is to determine both the optimal rotation of crops and corresponding irrigation policies. In this chapter analysis is confined to a perennial crop (citrus).

The model can generally be applied to analysis of alternative irrigation decisions in regard to quantity and quality of irrigation water at any time of the year. Due to limitations of empirical information, the specific model presented below relates to a subset of the overall decision options and is confined to the analysis of winter leaching policy. Quantity and quality of summer irrigation water is assumed to be constant. Water quality is varied parametrically and exogenously to the model.

A planning horizon of N stages or years is referred to with a single year being a basic unit in the sequence. Each year begins at the end of the rainy winter season. It is assumed that all rainfall is confined to the winter season, this being a good approximation to reality for the Mediterranean climate. The empirical analysis relates to a subhumid region in which need for supplementary irrigation during the winter season is negligible.

The following elements comprise the model: $\xi_i(n)$ denotes state i at the beginning of year n at the termination of the rainy season of the previous year. It is defined in terms of mean chloride concentration in the soil profile to a depth of 90 cm (meq Cl/l on a saturated paste basis). Fifty discrete states are distinguished ($i = 1, 2, \ldots, 50$).

d_i^k is the *k*th decision ($k = 1, 2, \ldots, K$) taken at state i in any year. Definition of the feasible decision set, from which the *k*th decision is selected, is independent of the state of the system and the year in the sequence. It is assumed that in all states and years the same decisions are *a priori* acceptable. The d_i^k decisions are defined in terms of quantity of water used to leach soil profile at the end of winter and before the summer irrigation season. The rationale for the definition of such decision options follows from a study by Bresler and Yaron (1972) that implies that the best timing for leaching is when the profile is saturated. Another reason for testing the validity of leaching at the end of the winter season is the relative abundance of water supply as compared with demand for water at this time, a fact reflected in the alternative cost of water.

Quality (salinity) of water is referred to as a parameter exogenous to the model with a different salinity level for each run. The quantity of water as applied during the irrigation season is assumed to be constant at the level practiced by farmers. The model can be adapted easily to accommodate the quality of water as an endogenous variable and varying summer irrigation levels as additional decision options.

$\theta_r(n)$ is the random *r*th level of rainfall in year n ($r = 1, 2, \ldots, R$), with probability $p[\theta_r(n)]$. $\left(\sum_{r=1}^{R} p[\theta_r(n)] = 1\right)$. It is assumed that the probability distribution of rainfall in year n is independent of the rainfall in year n. p_{ij}^k is the transformation probability of the system from state j following the *k*th decision with the corresponding transformation function being

$$(1) \qquad \xi_j(n-1) = t^k\left[\xi_i(n),\ \theta_r(n)\right].$$

For a given combination of state $\xi_i(n)$, decision d_i^k, and a rainfall level $\theta_r(n)$, there is a unique transformation from $\xi_i(n)$ to $\xi_j(n-1)$ with a probability of p_{ij}^k.

Due to the complexity of transformation (1) and the sequential nature of the salt leaching and accumulation process, this function was expressed in the empirical study by three successive auxiliary functions: (a) a salt leaching function expressing the effect of the leaching irrigation at the end of the rainy season,

(1.1) $$\xi_g^*(n) = g\left[\xi_i(n),\ d_i^k\right]$$

where ξ_g^* is the adjusted chloride concentration in the soil profile due to the leaching decision d_i^k; (b) a summer irrigation accumulation and/or leaching function,

(1.2) $$\xi_p^{**}(n) = g'\left[\xi_g^*(n)\right]$$

where ξ_p^{**} is the fall soil chloride concentration due to summer irrigation of a constant quantity; and (c) a winter rainfall leaching function,

(1.3) $$\xi_j(n-1) = g''\left[\xi_p^{**}(n),\ \theta_r(n)\right]$$

It is easily seen that integration of the functions (1.1) through (1.3) by successive substitutions gives the unique expression of the function (1).

f_i^k is the immediate net return in any year *n* derived from decision *k* at state *i.* It comprises (a) negative or zero returns, due to cost of leaching at the end of the winter and (b) a positive return due to the yield achieved in response to decision *k* at state *i.* This return takes into account the loss accrued to the yield due to salt concentration in the soil profile in the course of a single year.

$\Lambda_i(n)$ is the maximal expected value of cumulative net return at state *i* in year n computed in (2) below. The optimization problem is to maximize $\Lambda_i(N)$ for any initial state *i*, using the recursive equation

(2) $$\Lambda_i(n) = \max_k \left[f_i^k + \alpha \sum_{j=1}^{M} p_{ij}^k \Lambda_j(n-1) \right]$$

where $\alpha<1$ is a time discount factor, and $\Lambda_i(O) = 0$ for all *i*, for *N* large enough.

3. EMPIRICAL APPLICATION OF THE MODEL

In this section an application of the model to an economic evaluation of water quality for a selected case is presented. The case relates to citrus production—

Shamouti on sweet line rootstock—on medium texture soil (*SP* =37.5) in the Northern Coastal Plain of Israel, with mean annual rainfall of approximately 600 mm.

Economic data (yield, price, fixed and variable cost) were obtained from a citrus profitability survey in Israel (Farm Revenue Research Institute, 1971). Since these data are of local relevance, they are not presented here; however, note that the yield referred to was 12.4 tons per acre.

The physical relationships used are described as follows: a) End of winter salt leaching (1.1) was evaluated according to a model proposed by Bresler (1967). The soil profile was divided into three layers of 30 cm each, through which the body of water perforates. It was assumed that the soil profile was saturated to field capacity at the time of leaching, apparently a reasonable approximation for this time of year; and that leaching could be carried out only in the course of a single irrigation. b) Evaluation of salt accumulation and/or leaching during the summer irrigation season (1.2) was based on the law of mass conservation, whereby the amount of salt added to the soil profile by the body of water less the amount leaching out is equal to the terminal salinity content of the profile. The above basic identity can be written as

$$(3) \qquad Q_w Cl_w - \beta Q_w \frac{\left(\xi^* + \xi^{**}\right)}{2} = \left(\xi^{**} - \xi^*\right) \cdot V$$

where

Q_w = depth of irrigation applied (mm);

Cl_w = chloride concentration in the irrigation water (meq Cl/l);

β = fraction of irrigation water applied being leached out of the profile; and

V = amount of water contained in the relevant soil profile (mm), with ξ^* and ξ^{**} as undefined above.

The parameter β was evaluated to be 0.10. It was necessary to evaluate β due to the lack of detailed information on evapotranspiration and leaching in citrus irrigation. Relationship (4) is obtained from equation (3):

$$(4) \qquad \xi^{**} = Q_w Cl_w + \xi^* \left(V - \frac{\beta Q_w}{2}\right) \Big/ \left(V + \frac{\beta Q_w}{2}\right)$$

c) Leaching of the profile due to rainfall (1.3) was also computed on the basis of the above-mentioned relationship pertaining to the law of mass conservation. In this case, the salinity of rain was taken as being zero. Therefore identity (3) can be modified to

$$(5) \qquad \gamma R \frac{\left[\xi^{**}(n)+\xi(n-1)\right]}{2} = \left[\xi^{**}(n)-\xi(n-1)\right]\cdot V$$

where R is the amount of rainfall (mm) and γ ($0 < \gamma < 1$) is the fraction of the amount of rainfall effective in leaching. Other symbols were defined above. γ was evaluated on the basis of empirical observations, to equal 0.6. Use of equation (5) results in

$$(6) \qquad \xi(n-1) = \xi^{**}(n)\left(V-\gamma\frac{R}{2}\right)\bigg/\left(V+\gamma\frac{R}{2}\right).$$

Frequency distribution of discrete amounts of rainfall for the area of the empirical analysis was obtained from rainfall data accumulated over the last 35 years.

d) Evaluations of yield loss due to salinity were provided by soil scientists, working with data from local and other sources. It was assumed that average seasonal soil salinity of 7.5 meq Cl/l (at saturated extract) caused a yield loss of 10 percent, and soil salinity levels of 10, 15, 20, and 50 meq Cl/l caused a yield loss of 20 percent, 35 percent, and 50 percent, respectively.

Table 6.1. Annual expected steady state monetary losses due to soil and water salinity

Initial soil profile salinity meq Cl/l	Water salinity meq Cl/l						
	2	4	6	8	10	12	14
	IL/dunam per year[a]						
5	18	46	97	144	180	206	226
10	22	53	107	153	188	212	232
15	27	59	112	157	191	215	235
20	33	64	116	161	194	218	237
25	38	69	120	163	196	220	239

[a] 4.20 IL (Israeli Lira) = 1 US$

The model was applied to a sequence of 50 years ($N = 50$), long enough to set $\Lambda_i(O) = 0$ for all i, (the value of state i at the terminal year of the system) for $\alpha<1$. Rate of interest of 12 percent per year was assumed with $\alpha<1/1.12$. Different levels of irrigation water quality were applied for each run (2, 4, 6, 8, 10, 12, and 14, meq Cl/l).

Four leaching policies were tested—0 (no leaching), 100, 150, and 200 mm. Quality of water used for leaching was assumed to be the same as that of the summer irrigation water levels for 2-6 meq Cl/l and 6 meq Cl/l when summer irrigation water of higher salinity was applied.

It was assumed that the alternative cost of water used for leaching was lower than (0.1 IL/m^3) the "average" alternative cost, due to the particular season for leaching. Results of the analysis follow.

3.1. Winter Leaching

The results provide detailed information on the optimal leaching policy in terms of quantity to be used for each state of the soil profile at the end of the winter, quality of leaching water and quality of irrigation water in the following summer. It was found that convergence to a steady state policy occurs within three to five years. Optimal steady state policies were derived. For example, using summer irrigation water of quality 6 meq Cl/l and leaching water of the same quality, the optimal policy is to apply 200 mm for leaching whenever soil profile salinity is above 8 meq Cl/l.

3.2. Steady State Annual Monetary Losses

Expected steady state monetary losses on a yearly basis accrued to yields under optimal leaching policy, computed from the results are presented in Table 6.1.

3.3. Value Of Water Quality

The results presented in Table 6.1 served as a basis for the estimation of the value of water quality. For example, for an initial soil salinity of 5 meq Cl/l the value of one meq Cl/l in the water applied to one dunam of Shamouti on sweet lime (700 m^3) in the range 6-8 meq Cl/l is (144-97)/2 = 23.5 or 1/700 of this amount per one m^3, or IL. 0.03 approximately. Values for the entire range of quantities can be found in Table 6.2. The values for other initial salinity conditions are of similar magnitude

Table 6.2. Corresponding values for a range of water quality

Range of water quality	2-4	4-6	6-8	8-10	10-12	12-14
Value of one meq Cl/l	0.02	0.04	0.03	0.03	0.02	0.01

3.4. Estimated Value of Initial Soil Profile Salinity

As can be seen from Table 6.1, the effect of initial soil salinity on yield reduction is in the range of IL. 0.005-0.01 per meq Cl/l.

4. CONCLUSION

In this chapter a long-run approach to the economic evaluation of water quality in irrigation was described. While the emphasis of the chapter was on methodology, empirical estimates for one selected situation were presented. The estimates derived relied heavily on assumptions made with regard to salt leaching and accumulation and yield functions. These functions were formulated with the aid of soil scientists on the basis of a considerable amount of field data and may be considered good approximations of reality.

Decision variables in the analysis were restricted to quantities of water used for leaching, with the quality of water being an exogenous parameter. The model can accommodate other decision variables such as quality of the leaching water, quantity of irrigation water, and intervals between successive irrigations. With minor modifications the model can be applied to analyses of situations in which, beyond the possibility of varying water quantity and quality in irrigation, adjustments can be made in the cropping systems of the farms.

The economic significance of the model and the results, which may be derived through its use, can be considered on two levels: the micro farm level and the level of regional water resource development. Regarding the farm level, use of the model points to the optimal irrigation policy, with selection of decision variables depending on the particular situation of the farm. The most meaningful use of the model on this level would be with respect to farms, which have a water supply from different sources of varying qualities (e.g., farms in Beisan Valley in Israel). On farms without control of the quality of their water supply, the decision options are restricted to leaching decisions only. Such a case was referred to above.

Regarding the regional level, the model and the empirical estimates obtainable with its aid may be useful in the analysis of water resource development where the parameter of quality is taken into consideration. Water quality has gained increasing attention in recent years in numerous locations in the world (MacAvoy and Peterson, 1969; Pinock, 1969) due to the gradual process of pollution and quality deterioration of some major natural sources of water supply and the technological possibility of improving water quality by special treatments such as desalination. A comprehensive analysis of the value of water quality will be meaningful only if performed within the framework of an actual water demand and supply situation. While such an approach falls beyond the scope of this chapter, the possible magnitudes of assigning the "quality premium" to desalted water can be pointed out. Consider a hypothetical (but not unlikely) situation of citrus irrigation, 10 to 15 years in the future, in which the Marginal Value of Productivity (MVP) of a standard cubic meter of water with chloride content of 8 meq/l will be 0.20 IL/m^3. Assume that it will be mixed with one m^3 of desalted water at the ratio 1:1. In the result two m^3 of water will be contained with a chloride content of meq/l.

On the basis of the results presented here, if a quality premium of IL. 0.03 per 1 meq Cl/l water is assigned, the total value of the two cubic meters obtained after mixing will be IL. 0.64, leaving a net value of IL. 0.44 for the desalted cubic meter, i.e. more than twice the standard MVP.

The above only demonstrates a possible use of the model's results. A comprehensive analysis considering demand for water and the total water supply system should be undertaken to arrive at more specific conclusions about the value of water quality under any specific demand-supply situation. A major consideration in such an analysis will be the relative area of crops, such as citrus, sensitive to salinity.

REFERENCES

Bernstein, L. Salt tolerance of plants. *United States Department of Agriculture Agronomic Information Bulletin,* 283. 1964

Bernstein, L. Salt tolerance of fruit crops. *United States Department of Agriculture. Agronomic Information Bulletin,* 284. 1965.

Bierhuizen, J. F. Water quality and yield depression. Institute of Land and Water Management Research. Wageningen, Holland. *Technical Bulletin,* 61. 1969.

Bressler, E. A model for tracing salt distribution in the soil profile and estimating the efficient combination of water quality and quantity under varying field conditions. *Soil Science,* 104:227-223. October 1967.

Bressler, E. and D. Yaron. Soil water regime in economic evaluation of salinity in irrigation. *Water Resources Research,*. 8:791-800. August 1972.

DeForges, J. M. Research on the utilization of saline water for irrigation in Tunisia. *Nature and Resources*, 6:2-6. March 1970.

Farm Revenue Research Institute. Profitability of citrus industry 1968/9 (Mimeo). Farm Revenue Research Institute. Tel-Aviv, Israel. 1971. (Hebrew)

Howard, R. A. "Dynamic Programming and Markov Processes." Cambridge, MA. The M.I.T. Press. 1960.

MacAvoy, O. W. and D. F. Peterson. "Large Scale Desalting." Praeger Publishers. New York, NY. 1969.

Pinock, M. G. Assessing impacts of declining water quality on gross output of agriculture: A case study. *Water Resources Research,* 6:1-13. February 1969.

Shalhevet, J., P. Reiniger and D. Shimshi. Peanut response to uniform and non-uniform soil salinity. *Agronomy* Journal, 61:384-387. May-June 1969.

Timmons, J. F. and M. D. Dougal. Economics of water management. "Proceedings of International Conferences on Water for Peace." Superintendent of Government Documents. Washington, DC. 1967.

Yaron, D. Economic criteria for water resources development and allocation, II (Mimeo). The Hebrew University of Jerusalem. Rehovot, Israel. 1966.

Yaron, D. and E. Bressler. A model for the economic evaluation of water quality in irrigation. *Australian Journal of Agricultural. Economics,* 14:53-62. June 1970.

---

[1] J. Shalhevet and E. Bresler provided guidance in the formulation of the physical relationships and their quantitative evaluations. The study was funded by a grant from Resources for the Future Inc. and financial support from Mekorot, Israel Water Co.

7

A MODEL FOR OPTIMAL IRRIGATION SCHEDULING WITH SALINE WATER*

Dan Yaron
The Hebrew University of Jerusalem, Rehovot, Israel

Eshel Bresler and Hanoch Bielorai
The Volcani Center of Agricultural Research, Bet Dagan, Israel

Biniamin Harpinist
The Hebrew University of Jerusalem, Rehovot, Israel

1. INTRODUCTION

This chapter presents a model for optimal scheduling of irrigation with soil salinity parameters explicitly considered. Irrigation scheduling constitutes an essential component of proper irrigation management, particularly under conditions of scarcity of irrigation water and when excessive irrigation and runoff cause 'external damages,' as, for example, in the Colorado River Basin (Maletic, 1974; Young et al., 1974). Studies dealing with scheduling of irrigation (i.e., proper timing and quantities of water applied) have been published by Jensen and Heerman (1970), Kincaid and Heerman (1974), Dudley et al. (1971), Jensen (1972, 1977), Shalhevet et al. (1976), and others. However, most of the studies dealing with scheduling of irrigation have not taken parameters of salinity of the soil and of the irrigation water into account, the paper by Jensen (1977) being an exception.
The biological and agricultural aspects of salinity in irrigation have been extensively studied during the last few decades (e.g., Wadleigh and Ayers,

*Permission to publish this chapter was granted by the American Geophysical Union. The chapter was originally published under a similar title in *Water Resources Research*, 16:257-262, 1980.

1945; *U.S. Salinity Laboratory,* 1954; Hayward, 1954; Bernstein and Pearson, 1954; Bernstein, 1964, 1965; Bierhuizen, 1969; Bingham et al., 1969; Chapman et al., 1969; DeForges, 1970; Bernstein and Francois, 1973 and others). The physical relationships related to the management of irrigation with saline water and underlying economic analyses of irrigation with saline water have been previously discussed by Yaron and Bresler 1970; Bresler and Yaron 1972; Yaron et al., 1972; Bernstein and Francois 1973; Yaron 1974; Hanks and Andersen; 1979).

In economic analyses of irrigation with saline water, a distinction is useful between three types of models according to the range of time to which they refer: a) 'Short run' models refer to the relationships confined to 1 year or a single irrigation season. The short run models consider the initial salinity of the soil profile in the root zone at the beginning of the irrigation season as given and, for each initial state, analyze the optimal combination between water quantity and quality. Short run models, however, do not take into account the effects of accumulation of salt over time (e.g., Yaron and Bresler, 1970; Bresler and Yaron, 1972; Moore et al. 1974).

b) 'Long run' models do account for the effects of salt accumulation in the soil profile over time. Such models comprise a succession of short run processes, the initial conditions of which are affected by salt accumulation in previous periods. The irrigation decisions over a single season take into account the resulting terminal conditions due to each alternative considered and their effects in the succeeding periods (Yaron and Olian, 1973).

c) 'Extended long run' models take into account both the salt accumulation in the root zone soil profile as well as its accumulation in the underground reservoirs (e.g., Hanks and Andersen, 1979). Note that under conditions in which runoff irrigation water reaches the water sources used for irrigation in a relatively short time, the distinction between the long run and the extended long run model may not be applicable. Such conditions exist, for example, in the irrigated areas of the Colorado River Basin in the United States or in the coastal plain of Israel.

The model presented in this chapter is one of the short run models. Its analysis provides answers to two questions arising under conditions of irrigation with saline water: (1) Given the initial soil salinity, should a pre-planting leaching be applied and, if so, at what quantity? (2) What is the optimal irrigation scheduling, i.e., the combination of quantities and timing during the entire irrigation season? This chapter outlines the essentials of the model, presents its details and the specification of its elements, and demonstrates an application of the model to optimal scheduling of irrigation with saline water to sorghum crop. A short summary evaluates the applicability or the model and suggests possible modifications and extensions.

2. OUTLINE OF THE MODEL

A dynamic programming model designed to determine the optimal irrigation scheduling was presented by Hall and Butcher (1968). In their model, however, the salinity dimension was not considered. The dynamic programming approach is extended in our model to account for the response of crop yields to soil moisture as well as to soil salinity.

The system underlying the model is characterized by two discrete state variables: θ_t^p is the soil moisture level p ($p = 1, 2, \ldots, P$) at the beginning of day t, and c_t^s is the salinity level s ($s = 1, 2, \ldots, S$) of the soil solution at the beginning of day t, which together determine the total soil potential ψ (Bresler and Yaron, 1972).

$$\text{(1)} \qquad \psi_t = \psi(\theta_t, C_t)$$

The growing season is subdivided into J subperiods in accordance with the stages of growth of the crop. The yield of the crop is expressed by a function of the following type (Yaron et al., 1973):

$$\text{(2)} \qquad Y = A\prod_{j=1}^{J}(F_j)^{x_j}$$

where

Y = yield, kg/ha;
A = maximal yield obtained when all $x_i = 0$;
x_j = number of 'critical days' in subperiod j (a critical day is defined as one during which the total soil potential ψ exceeds a critical level);
F_j = coefficient of yield reduction per each critical day during subperiod j, $0 < F_j < 1$.

The specification of the independent variables in (2), in terms of the number of critical days, originates in crop response studies conducted by Yaron et al. (1972) and Yaron et al. (1973). However, other definitions of the independent variables are also possible. Moreover, alternative formulations of the crop response function may be introduced, with one example of an alternative being the linear function

(3) $$Y = A - \sum_{j=1}^{J} F_j^* x_j$$

with F_j^* being the reduction in yield (in kilograms per hectare) per each critical day in subperiod j.

Note that the above specifications of the response functions (2) and (3) implicitly assume that the current and future response of a crop to irrigation is not affected by the number of critical days in previous subperiods and the previous history of the crop. Specifications applying a similar assumption have been adopted by others (e.g. Hall and Butcher, 1968; Hanks, 1974; Minhas et al., 1974). Additional empirical data is needed in order to test alternative specifications and hypotheses regarding subperiod interactions.

On each day a decision regarding whether to irrigate and how much water has to be applied is considered. The decision is taken after the state of the system $\left(\theta_t^p,\ c_t^s\right)$ is determined. Depending on the state and the decision Q_t^k taken ($k = 1, 2, ..., K$) an 'immediate loss' $f_t^k\left(Q_t^k \middle|\ \theta_t^p,\ c_t^s\right)$ is incurred. This function comprises the cost of irrigation if applied, or the loss due to the potential decrement in the ultimate yield on day t, if irrigation is not applied, and the total soil potential by the end of day (ψ_{t+1}) rises above a prescribed critical value.

The core of the model is the following recursive relationship:

(4) $$\Lambda_t\left(\theta_t^p,\ c_t^s\right) = \max_{Q_t^k}\left[f_t^k\left(Q_t^k,\ \theta_t^p,\ c_t^s\right) + \Lambda_{t+1}\left(\theta_{t+1}^{p*},\ c_{t+1}^{s*}\right)\right]$$

$t = 0, 1, 2, \ldots, T\text{-}1$ with $f_t^k() \leq 0$ and

(5) $$\begin{aligned} \theta_{t+1}^{p*} &= g_t\left(\theta_t^p,\ Q_t^k \middle|\ G\right) \\ c_{t+1}^{s*} &= h\left(\theta_t^p,\ c_t^s,\ Q_t^k \middle|\ G\right) \end{aligned}$$

(6) $$\Lambda_T\left(\theta_T^p,\ c_T^s\right) = A \cdot P_y - FC$$

for all p and s, where P_y is the price per yield unit (net or harvest cost) and FC is the fixed cost per land unit area, with cost or irrigation and harvest excluded. Note that FC (a constant) can be ignored in the recursive maximization process.

The functions g_t and h in (5) are the transformation functions of soil moisture and soil salinity from day t to $(t + 1)$; θ, c, and Q were previously defined, and G represents all other factors, considered as constant. The index

t stands for the number or days from the beginning or the growing season, with $t = 0$ being the planting day and $t = T$ the end of the growing season. Note that the direction of change in t is opposite to the conventional notation in dynamic programming.

The objective is to maximize the cumulative net income $\Lambda_0\left(\theta_0^p,\ c_0^s\right)$ for every p and s, subject to (5) and (6), by applying a dynamic programming backward induction procedure to (4) for $t = T, T - 1, T - 2$, and so on, with $t = 0$ denoting the beginning of the growing season.

3. A DETAILED SPECIFICATION OF THE MODEL

The model has been applied to the analysis of the optimal irrigation with saline water or sorghum in the Gilat area (in northwestern Negev, Israel) with saline water. The sorghum crop has been selected for the empirical application in view of the relative availability of the information needed, derived from sorghum irrigation experiments (Bielorai et al, 1964). Each element of the model is presented in this section both in general and in specific terms.

3.1. The Growing Season

The growing season with T days is subdivided into four ($J = 4$) subperiods according to the stages or growth of the crop. In period j there are T_j days, with $\sum_{j=1}^{J} T_j = T$ (108 days). The periods distinguished were from sowing to tillering ($j = 1$), from tillering to heading ($j = 2$), from heading to the milking stage ($j = 3$) and from the milking stage to the mature stage ($j = 4$).

There is no rainfall during the growing season, a situation typical for summer crops in Mediterranean climate and many other arid and semiarid areas.

3.2. The Yield Function

The basis for the yield function applied was an empirical estimate of a response function of grain sorghum to soil moisture, derived from the Gilat experimental data and modified to incorporate salinity. The following function was assumed (Bielorai and Yaron, 1978):

(7) $$Y = 9000(0.995)^{x_1}(0.98)^{x_2+x_3}(0.995)^{x_4}$$

where Y is the yield of grain sorghum (in kilograms per hectare) and x_j is the number of critical days in subperiod j, with a critical day being defined as one in which the total soil potential exceeds 2 bars ($\psi > 2$). This function implies that the maximal yield is 9000 kg/ha and that each critical day in the non-reproductive period reduces the yield by 0.5 percent and, in the reproductive period, by 2 percent.

It should be noted that while the number of critical days, x_j in (7), are defined here in terms of total soil water potential (equal to the sum of matric and osmotic potentials), the variables x_j ($j = 1, 2, ..., 4$) in the original estimation of the response function (7) (Bielorai and Yaron, 1978) were in terms of available soil water content. These critical values (x_j) were transformed into soil matric potential using the soil water retentivity curve of Gilat soil. The corresponding total soil water potential ψ was then evaluated to account for the sum of matric and osmotic potentials.

3.3. State Variables

The state variables were specified in terms of θ^p, the soil moisture, on weight percentage basis, averaged over the main root zone (0-90 cm) (47 discrete states of soil moisture were distinguished with the range being between 18.5 percent (field capacity) and 6.75 percent (permanent wilting point), with increments of 0.25 percent between states), and c^s, the chloride concentration of the saturated paste soil solution in the root zone (in milli-equivalents of chloride per liter) (30 states were distinguished, 1, 2, ..., 30 meq Cl/l. The overall number of the states was 1410 ($= 47 \times 30$).

3.4. Decision Variables

The decision variables Q_t^k were the time and the depth of irrigation water applied on day t, with the alternatives considered being 0, 40, 80, 120 and 160 mm per one irrigation application. Irrigation was assumed to be applicable at intervals of 6 days, namely, on $t = 1, 7, ..., 103$. Accordingly, regarding days other than 1, 7, ..., 103, only one value of the decision variable $Q_t^k = 0$ was allowed; the implications of the above with respect to the transformation functions (5) are discussed in the following section.

The salinity of the irrigation water was considered as an exogenous factor and constant throughout the growing season in each run, an assumption, which complies with the realistic conditions in the area. The salinity levels were changed parametrically over the values 2.5, 5, 7.5, 10 and 15 meq Cl/l. Note that it is easy to incorporate the variable salinity of the irrigation water into the model. However, this implies that the decision maker should have control over the salinity of the irrigation water used.

3.5. Transformation Functions From One State to Another

Models of different type, precision, and complexity can be used to simulate the soil water content and salinity in order to be applied in the salinity and water content transformation functions (e.g., Bresler, 1967, 1973; Bresler and Hanks, 1969; Childs and Hanks, 1975; Hanks. 1974; Feddes et al., 1974; Neuman et al., 1975; Nimah and Hanks, 1973). For simplicity and demonstration purposes of the capability of our model the less complicated water and salt flow models were chosen. These simple models, however, have been proven to be reasonably good tools for prediction purposes (Yaron et al., 1973; Yaron, 1974; Shimshi et al., 1975; Yaron and Bresler, 1970).

The soil water transformation function

$$\theta_{t+1}^{p*} = g_t\left(\theta_t^p,\ Q_t^k \middle|\ G\right)$$

was expressed by using a soil moisture budgeting simulation routine described by Yaron et al. (1973) and Shimshi et al. (1975). In essence this routine assumes that on any day on which irrigation is applied, the soil water in the root zone is raised in accordance with the quantity of water applied and the soil moisture deficit prior to irrigation. On any day without irrigation the soil water is depleted by the evapotranspiration ET_t expressed by the following functional relationship:

(8) $$ET_t = a_t + b_t\theta_t(= \theta_t - \theta_{t+1})$$

The parameters a_t and b_t were estimated from the experimental data and their values are below:

Table 7.1. The estimated parameters (a_t, b_t) of the Gilat experiments[a]

t	a_t	b_t
1-30	-0.175	0.02612
31-60	-0.375	0.05597
61-108	-0.475	0.07089

[a]The above parameters were estimated from an irrigation experiment in which soil salinity was relatively low and was not measured. Accordingly, they are valid for low soil salinity situations in which ET is not affected by salinity and provide only an approximation to the true evapotranspiration function otherwise.

The evapotranspiration model is deterministic, exploiting the relative predictability of evapotranspiration in the region; however, the model could be easily extended to incorporate stochastic soil moisture transformation functions.

The soil salinity transformation function

$$c_{t+1}^{s*} = g\left(\theta_t^p,\ c_t^s,\ Q_t^k \middle|\ G\right)$$

was specified for $Q_t^k > 0$ (days on which irrigation was applied) using the dynamic salt balance equation (9) presented and discussed by Bresler (1967) and empirically tested by Yaron and Bresler (1970) and Yaron (1974):

(9) $$Q_t^k \cdot CLW_t - (Q_t^k - DEF_t) \cdot \frac{c_t + c_{t+1}}{2} = (c_{t+1} - c_t) \cdot V$$

where

Q^k = depth k of irrigation water applied, mm;
CLW = chloride concentration in the irrigation water, meq Cl/1
DEF = soil moisture deficit in the root zone layer, mm;
V = depth of water contained in the root zone during leaching, mm;
and c_t is as previously defined. Solving for c_{t+1}, we get

(10) $$c_{t+1} = \left(2Q_t \cdot CLW_t - c_t\left(Q_t - DEF_t - 2V\right)\right) / \left(Q_t - DEF_t + 2V\right)$$

The values of *DEF* were obtained from the soil moisture updating routine (equation (8)). For days with $Q_t^k = 0$, i.e., no irrigation, $c_{t+1} = c_t$.

3.6. Auxiliary Functions

As previously stated, the yield was assumed to be affected by the number of critical days in each of the *J* periods. To compute ψ_t for any state $\left(\theta_t^p,\ c_t^s\right)$, the following auxiliary functions were used (Bresler and Yaron, 1972):

(11) $$M_t^p = \tau(\theta_t^p)$$

(12) $$O_t^{ps} = \phi\left(\theta_t^p,\ c_t^s\right)$$

(13) $$\psi_t^{ps} = M_t^p + O_t^{ps}$$

with *M* and *O* being the matric and osmotic soil water potential, respectively.

The matric soil potential as a function of soil moisture (11) was measured by Bielorai and Levy (1971) on disturbed soil samples by the conventional pressure plate technique. As M_t is not a single-valued function of θ_t because of hysteresis, the relevant part of the M - θ relationship was taken to be the drying part of the hysteresis loop in the water retention curve. The osmotic potential of the soil solution was expressed and computed using the following relationship (Bresler and Yaron, 1972):

$$(14) \quad c_t^{act} = c_t \frac{WSP}{\theta_t}$$

$$(15) \quad EC_t = \alpha + \beta c_t^{act}$$

$$(16) \quad O_t = \gamma \cdot EC_t^{\delta}$$

with

c_t^{act} = chloride concentration in the soil solution at the actual soil moisture w_o meq Cl/l;
WSP = soil moisture content at saturated paste;
EC_t = electrical conductivity of the soil solution, mm ha/cm.

The values of the parameters α and β were estimated from empirical data to be 0.8 and 0.109, respectively. The functional relationship (16) was derived by fitting an equation to Figure 6 of U.S. Salinity Laboratory (1954, p. 15), with the parameters γ and δ found to be 0.321 and 1.0812, respectively.

3.7. The Immediate Loss Function

The immediate loss function comprises cost of irrigation $(Q_t^k \cdot P_w)$ with Q_t^k being the net quantity of irrigation water, P_w the cost per water unit (0.55 IL/net m^3) (IL = Israeli Lira), and the monetary value of the potential loss of yield L_t. Quantity of water was defined in terms of 'net' cubic meter to avoid adjustments needed for different levels of irrigation efficiency.

To compute L_t, the potential yield level was updated and stored for any state $(\theta_t^p,\ c_t^s)$ in parallel to the dynamic programming backward recursive process, beginning with the maximal potential yield $Y_T = A$ for $t = T$.

The updated yield as a function of $(\theta_t^p,\ c_t^s)$ and Q_t^k is defined as

(17)
$$Y_t^{ps}\left(\theta_t^p,\ c_t^s,\ Q_t^k\right)=Y_{t+1}^{p*s*}\cdot\left(1-F_t\right) \qquad \psi_{t+1}^{p*s*}>2$$
$$Y_t^{ps}\left(\theta_t^p,\ c_t^s,\ Q_t^k\right)=Y_{t+1}^{p*s*} \qquad \psi_{t+1}^{p*s*}\leq 2$$

with F_t being the subseasonal coefficients of the yield function (7). Given θ_t^p, c_t^s and Q_t^k, ψ_{t+1}^{p*s*} was computed by applying the soil moisture updating routine and functions (10)-(16).

Similarly, $L_t(\ \ )$ and the immediate loss function $f_t^k(\ \ )$ are defined and computed as

(18)
$$L_t\left(\theta_t^p,\ c_t^s,\ Q_t^k\right)=Y_{t+1}^{p*s*}\cdot P_y\left(1-F_t\right) \qquad \psi_{t+1}^{p*s*}>2$$
$$L_t\left(\theta_t^p,\ c_t^s,\ Q_t^k\right)=0 \qquad \psi_{t+1}^{p*s*}\leq 2$$

with P_y being the net price of sorghum grain, excluding harvest cost (900 IL/ton); and

(19) $$f_t^k\left(\theta_t^p,\ c_t^s,\ Q_t^k\right)=Q_t^k\cdot P_w+L_t^k\left(Q_t^k,\ O_t^p,\ c_t^s\right)$$

Note that in the recursive relationship (4)

$$f_t^k\leq 0 \qquad \Lambda_{t+1}>0.$$

With the elements of the model specified by (7)-(19), the objective function is maximization of $\Lambda_o\left(\theta_o^p,\ c_o^s\right)$ for every p and s subject to the recursive relationship (4), the transformation functions (5), and the maximum terminal income (6).

As previously stated, irrigation was assumed to be applicable at intervals of 6 days, with a full dynamic programming stage being 6 days long. Accordingly, computer calculations were classified into two subsets (1) for days with applicable irrigation and (2) for the other days. This distinction involves a considerable saving in computer time.

4. AN APPLICATION OF THE MODEL TO SELECTED SITUATIONS: RESULTS AND DISCUSSION

The model has been applied to the analysis of optimal irrigation policy of grain sorghum under several situations in the Gilat area. The adjustments in the optimal irrigation policy and the changes in the yield and the income per land unit area were studied with regard to two variables: the salinity of the irrigation water throughout the irrigation season and the initial soil salinity at the beginning of the growing season ($t = 0$). Selected results of these analyses are presented in the various parts of figure 7.1. More details and analyses of additional situations are given by Harpinist (1979).

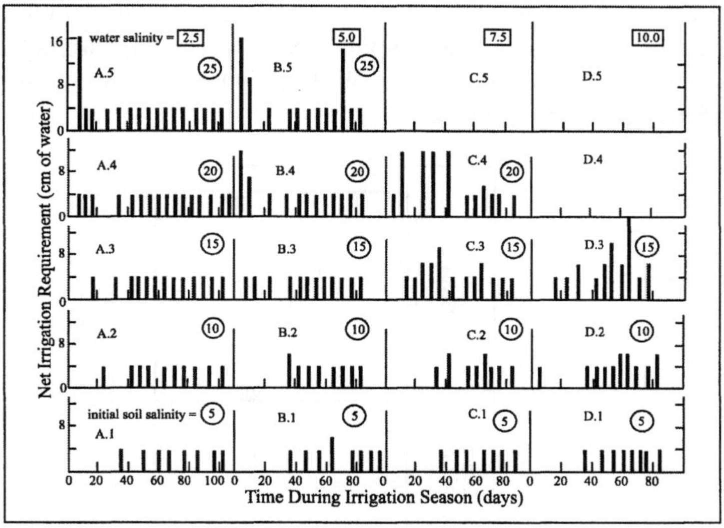

Figure 7.1. Net irrigation water requirement as a function of time during the irrigation season for five levels of initial soil salinity and four levels of salinity of irrigation water

The results of Figure 7.1 suggest several policy rules for irrigation with saline water under different initial salinity regimes: a) Generally, frequent applications of small quantities of water are preferable to application of large quantities at extended intervals. b) Under relatively high saline conditions (i.e., the use of high values of either initial soil salinity or water salinity, or both) extra irrigation water for leaching is generally justified at the beginning of the growing season (for example, situations A.5 and B.4, in Figure 7.1) or in the middle of the irrigation season (C.3 and D.3). Some combinations of the above are also recommended (B.5 and C.4). c) Under relatively low-saline conditions it is worthwhile to extend irrigation over a longer period, as compared to the high salinity affected situations (for example, B.1 versus B.2). The rationale underlying this rule is that when the yield potential is high, it pays to preserve the yield potential by extending the application of irrigation. d) Under the most saline conditions referred to in the analysis it is not worthwhile to irrigate at all.

Note that rules a) and b) have been recently set forth in general terms (e.g., Goldberg and Shmueli, 1969; Goldberg et al., 1971; Bernstein and Francois, 1973). The present model quantifies these rules for specific situations.

Figure 7.2 presents net income, annual quantities of irrigation water and sorghum yield as a function of initial soil salinity and salinity of irrigation water. Decreasing income when both water and initial soil salinity rise is clearly seen in the upper part of Figure 7.2. The results presented in Figure 7.2 point, as well, to a general tendency to preserve high-yield levels at the assumed cost of intensive water inputs. This tendency apparently reflects the water/sorghum price ratio assumed. With a higher water/sorghum price ratio this tendency could be reversed.

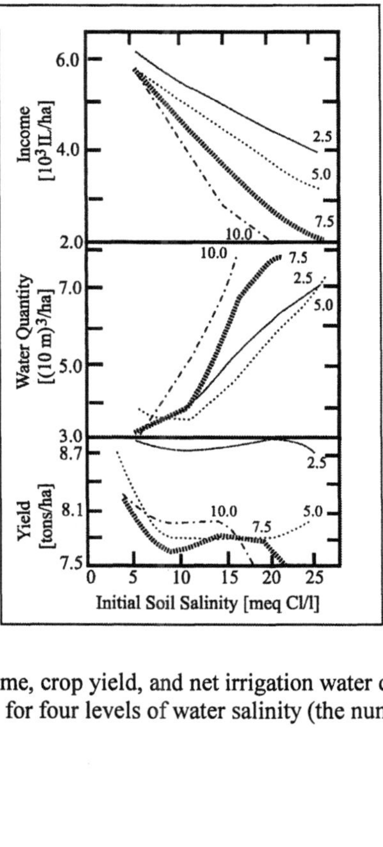

Figure 7.2. Net income, crop yield, and net irrigation water quantities, as a function of initial soil salinity for four levels of water salinity (the numbers labeling the lines)

5. CONCLUSION

A dynamic programming model for optimal scheduling or irrigation with saline water has been presented. The elements of the model have been specified in detail with respect to irrigation of sorghum in the Gilat semiarid area of Israel. The model has been applied to the analysis of several specific situations.

In conclusion, it seems that the model might be useful in two major applications. First, in testing and screening irrigation policies for detailed examination by field experiments. Since computer simulation of the water-soil-crop system is considerably cheaper than field experiments, it can be used as a means for screening irrigation policies to help decide upon more expensive and more reliable field experiments to follow. Note that 6 CPU min on a 6,600 CDC computer are required for running the dynamic programming model. Second, the model can be used for detailed analyses of optimal irrigation with saline water under well-defined situations without field experiments. For such application however, refinement and calibration of the model for the specific situations is needed. A useful refinement may be to refer to distinct soil layers (possibly 30 cm of depth each) and to compute the moisture and salinity for each layer. Furthermore, empirical estimates of the evapotranspiration and the yield functions could be improved by reference to a wider database, including empirical information on the variation of both soil moisture and soil salinity.

REFERENCES

Bernstein, L. Salt tolerance of plants. *United States Department of Agriculture Information Bulletin,* 283. 1964.

Bernstein, L. Salt tolerance of fruit crops. *United States Department of Agriculture Information Bulletin,* 284. 1965.

Bernstein, L. and L. E. Francois. Comparisons of drip, furrow and sprinkler irrigation. *Soil Science,* 110:73-85. 1973.

Bernstein, L. and G. A. Pearson. Influence of integrated moisture stress achieved by varying the osmotic pressure of culture solutions on growth of tomato and pepper plants. *Soil Science,* 77:355-368. 1954.

Bielorai, H. and I. Levy. Irrigation regimes in a semi-arid area and their effects on grape fruit yield, water use and salinity. *Israeli Journal of Agricultural Science,* 21:3-13. 1971.

Bielorai, H. and D. Yaron. Methodology and empirical estimates of the response function of sorghum to irrigation and soil moisture. *Water Resources Bulletin,* 14:966-977. 1978.

Bielorai, H., I. Arnon, A. Blum, Y. Elkana and A. Reiss. The effects of irrigation and inter-row spacing in grain sorghum production. *Israeli Journal of Agricultural Resources,* 14:227-236. 1964.

Bierhuizen, I. F. Water quality and yield depression. *Technical Bulletin,* 61. Institute of Land and Water Management Research. Wageningen, Netherlands. 1969.

Bingham, F. T., L. H. Stobry and H. D. Chapman. Effects of variable water quality on Valencia tree performance. *Proceedings of the First International Citrus Symposium,* 3:1803-1809. 1969.

Bresler, E. A model for tracing salt distribution in the soil profile and estimating the efficient combination of water quality under varying field conditions. *Soil Science,* 104:227-233. 1967.

Bresler, E. Simultaneous transport of solute and water under transient unsaturated flow conditions. *Water Resources Research,* 9:975-986. 1973.

Bresler, E. and R. I. Hanks. Numerical method for estimating simultaneous flow of water and salt in unsaturated soils. *Soil Scence Society of America Proceedings,* 33:827-832. 1969.

Bresler, E. and D. Yaron. Soil water regime in economic evaluation of salinity in irrigation. *Water Resources Res*earch, 8:791-800. 1972.

Chapman, H. D., J. Harrietan and D. S. Rayner. Effects of variable maintained chloride levels on orange growth, yield and leaf composition. *Proceedings of the First International Citrus Symposium,* 3:1811-1817. 1969.

Childs, S. W. and R. J. Hanks. Model of soil salinity effects on crop growth. *Soil Science Society of America Proceedings,* 39:617-622. 1975.

DeForges, J. M. Research on the utilization of saline water for irrigation in Tunisia. *Natural Resources,* 6:2-6. 1970.

Dudley, N., D. Howell and W. Musgrave. Optimal intraseasonal irrigation allocation. *Water Resources Research,* 7:770-788. 1971.

Feddes, R. A., E. Bresler and S. P. Neuman. Field test of a modified numerical model for water uptake by root system. *Water Resources Research,* 10:1199-1206. 1974.

Goldberg, D. and M. Shmueli. A method for increased agricultural production under conditions of saline water and adverse soils. *Water Australia,* 6-10. April-May 1969.

Goldberg, D. D., M. Rinot and N. Kary. Effect of trickle irrigation intervals on distribution and utilization of soil moisture in a vineyard. *Soil Science Society of America Proceeding,* 35:127-130. 1971.

Hall, W. A. and W. S. Butcher. Optimal timing of irrigation. *Journal of Irrigation and Drainage Division American. Society of Civil Engineers,* 94 (IR2):267-275. 1968.

Hanks, R. J. Model for predicting plant yield as influenced by water use. *Agronomy Journal,* 66:660-665. 1974.

Hanks, R. J. and J. C. Andersen. Physical and economic evaluation of irrigation return flow and salinity on a farm in: "Salinity in Irrigation and Water Resources" (D. Yaron, ed.), Marcel Dekker. New York, NY. 1979.

Harpinist, B. Application or dynamic programming to the analysis of optimal timing of irrigation with saline water (MSc Thesis). The Hebrew University. Jerusalem, Israel. 1979. (Hebrew)

Hayward, H. E. Plant growth under saline conditions. *Arid Zone Research,* 4:37-71. 1954.

Jensen, M. E. Programming irrigation for greater efficiency in: "Optimizing the Soil Physical Environment Toward Greater Crop Yields" (D. Hillel, ed.). Academic Press. New York. 133-161. 1972.

Jensen, M. E. Scientific irrigation scheduling for salinity control of irrigation return flow. Paper presented at the National Conference on Irrigation Return Flow Quality Management. Ft. Collins, CO. 1977.

Jensen, M. E. and D. F. Heerman. Meteorological approaches to irrigation scheduling. Paper presented at the National Irrigation Symposium. American Society of Agricultural Engineers. Lincoln, NA. November 1970.

Kincaid, D. C. and D. F. Heerman. Scheduling irrigation using a programmable calculator. Rep. ARS-NC-12. United States Department of Agriculture. Washington, DC. February 1974.

Malelic, J. T. Current approaches to salinity management in the Colorado River Basin. Salinity in water resources. "Proceedings of the 15th Annual Western Resources Conference" (J. E. Flack and C. W. Howe, eds.). Boulder, CO. 1974.

Minhas, B. S., K. S. Parikh and T. N. Srinivasan. Toward the structure of a production function of wheat yields with dated inputs of irrigation water. *Water Resources Research,* 10:383-393. 1974.

Moore, C. V., J. H. Snyder and P. Sun. Effects of Colorado River water quality and supply on irrigated agriculture. *Water Resources Research,* 10:137-144. 1974.

Neuman, S. P., R. A. Feddes and E. Bresler. Finite element analysis of two-dimensional flow in soils considering water uptake by roots, I. Theory. *Soil Science Society of America Proceedings*, 39 (2):224-230. 1975.

Nimah, M. and R. J. Hanks. Model for estimating soil, water, plant, and atmosphere interrelations, I. Description and sensitivity. *Soil Science Society of America Proceedings,* 37:522-527. 1973.

Shalhevet, Y., A. Mantell. H. Bielorai and D. Shimshi. Irrigation of field and orchard crops under semi-arid conditions. Publication of International Irrigation Information Center. Bet-Dagan, Israel. 1976.

Shimshi, D., D. Yaron, E. Bresler, M. Weisbrod and G. Strateener. Simulation model for evapotranspiration of wheat: Empirical approach. *Journal of Irrigation and Drainage Division American Society of Civil Engineers,* 101:1-12. 1975.

U. S. Salinity Laboratory, Diagnosis and Improvement of Saline and Alkali Soils. Handbook 60. United States Department of Agriculture. Washington, DC. 1954.

Wadleigh, C. H. and A. D. Ayers. Growth and biochemical composition of bean plants as conditioned by soil moisture tension and soil concentration. *Plant Physiology*, 20:106-132. 1945.

Yaron, D. Economic analysis of optimal use of saline water in irrigation and the evaluation of water quality in: "Salinity in Water Resources: Proceedings of the 15th Annual Western Resources Conference" (J. E. Flack and C. W. Howe, eds.). Boulder, CO. 1974.

Yaron, D. and E. Bresler. A model for the economic evaluation of water quality in irrigation, *Australian Journal of Agricultural Economics,* 14:53-62. 1970.

Yaron, D. and A. Olian. Application of dynamic programming in Markov chains in the evaluation of water quality in irrigation. *American Journal of Agricultural Economics,* 55:467-471. 1973.

Yaron, D., H. Bielorai, J. Shalhevet and Y. Gavish. Estimation procedures for response functions of crops to soil water content and salinity. *Water Resources Research,* 8:291-300. 1972.

Yaron, D., G. Strateener, D. Shimshi and M. Weisbrod. Wheat response to soil moisture and the optimal irrigation policy under conditions of unstable rainfall. *Water Resources Research,* 9:1145-1154. 1973.

Young, R. A., W. T. Franklin and K. C. Nobe. Evaluating agricultural effects of salinity abatement projects in the Colorado River Basin agronomic and economic considerations in: "Salinity in Water Resources: Proceedings of the 15th Annual Western Resources Conference" (J. E. Flack and C. W. Howe, eds.). Boulder, CO. 1974.

8

OPTIMAL ALLOCATION OF FARM IRRIGATION WATER DURING PEAK SEASONS*

Dan Yaron
The Hebrew University of Jerusalem, Rehovot, Israel

Ariel Dinar
The Hebrew University of Jerusalem, and
The Israeli Ministry of Agriculture, Rehovot, Israel

1. INTRODUCTION

Allocation of irrigation water to competing crops or to different plots of the same crop is an important issue when water is scarce. This is especially true during peak seasons of demand for irrigation water. Water allocation among competing farm uses has been treated in numerous linear programming studies (Yaron). However the early linear programming (LP) studies have incorporated crop response functions to irrigation and soil moisture only indirectly, usually without rigorous analysis.

On the other hand, the literature contains some dynamic programming models intended to optimize irrigation scheduling (Hall and Butcher, 1968; Howell et al., 1975; Yaron et al., 1980). The dynamic programming models deal with a single crop, ignoring overall farm restrictions. They assume that water cost is exogenously predetermined and ignore shadow prices of water generated by the farm's production and irrigation program. Exceptions to this dichotomy are (a) Dudley et al. (1971), who applied a simulation model to choosing optimal acreage choice for irrigated crops with explicit crop response functions incorporated; and (b) Blank (1975), who applied linear pro-

*Permission to publish this chapter was granted by the American Agricultural Economics Association. The chapter was originally published under a similar title, in *the American Journal of Agricultural Economics*, 64(4):681-689, 1982

gramming to a multiple crop problem with several irrigation activities and response functions for selected crops.

This paper presents a systems approach to intrafarm water allocation and irrigation scheduling for major crops. The overall system contains two interrelated subsystems:

Subsystem I is an LP model intended to maximize the farm's income subject to constraints with given technology. Water supply restrictions are expressed in some detail. The peak season is subdivided into operational time units. Several irrigation alternatives for major crops during the peak season are incorporated. Because the number of sound *a priori* alternatives is large, irrigation scheduling activities are gradually incorporated into the LP model, using the "generalized Wolfe" LP algorithm (Dantzig, 1963, pp. 433-40).

Subsystem II is a dynamic programming (DP) model intended to generate new irrigation scheduling activities with shadow prices of water given by the LP solutions. The dynamic programming model involves two subroutines: (a) a soil moisture model which expresses the relationship between irrigation decision variables (timing, quantity) and the soil moisture fluctuations, given atmospheric conditions and other factors, and (b) a crop response model which relates variations in soil moisture to crop yields.

The analysis starts with the solution of an LP problem, given an initial vector of water prices. This yields a vector of shadow prices of water. These are incorporated into Subsystem II, generating new irrigation activities, which, if they exist, may improve the current LP solution. The solution of the overall system follows a loop procedure until the optimal solution is achieved. Convergence to optimality is ensured.

2. SYSTEM I

This is a "generalized" LP model (Dantzig) of the farm, including several predetermined irrigation-scheduling activities and a link with a DP model intended to generate new activities.

Consider the following LP problem:

maximize z

(1) $z = c'x$

subject to

$$q'x \leq b_T$$
$$Ax \leq b$$
(2) $$Gx \leq d$$
$$x \geq 0,$$

where c' is the $(1 \times n)$ vector of net income parameters; x, vector of activity levels $(n \times 1)$; b_T, total amount of water available for the irrigation season; b, vector $(m \times 1)$ of sub-seasonal water restrictions (on a ten day period basis, hereafter referred to as a "decade"); d, $(m * \times 1)$ vector of other constraints; and q', A, and G are the corresponding vector and matrices of input and output coefficients.

The problem involves irrigation activities conducted according to predetermined schedules with the initial LP problem containing conventional farm activities.

Assume that an optimal solution to (1) and (2) has been obtained, yielding a vector of shadow prices $\left[u_T^0,\ u^0,\ v^0\right]$, with $\left[u_T^0,\ u^0\right]$ relating to water (seasonal, total, and decade restrictions) and v^0 to other farm restrictions. If a new activity $\left[q^s,\ a'^s,\ g'^s\right]$ can be found such that

(3) $$q^s u_T^0 + a'^s u^0 + g'^s v^0 < c^s,$$

its incorporation into (1) and (2) will improve the current optimal solution. The left side of (3) is the alternative cost of the new activity, s and c^s is its income defined as

$$c^s = Y^s \cdot P_y - VC^s$$

is crop yield per activity unit (kilograms per hectare), P_y is yield price per kilogram, net of harvest cost, and VC^s is variable cost of purchased inputs per new activity unit.

To find an activity, which fulfills condition (3), we formulate the following problem: minimize $\left(q^s u_T^0 + a'^s u^0 + g'^s v^0 - c^s\right)$, or alternatively, maximize z^*

(4) $$z^* = Y^s \cdot P_y - VC^s - q^s u_T^0 - a'^s u^0 - g'^s v^0,$$

subject to

$$(5) \qquad \sum_{i=1}^{m} a_i^s = q^s,$$

where a_i^s is water input per unit of new activity in decade i (a non-negative, discrete value) and q^s is the seasonal total water input per unit of new activity.

Note that q^s is a decision variable, which can take a limited number (R) of discrete values specifying the water/land ratio for the new activity. Accordingly, there are R parametric variants of equations (4) and (5), each with a distinct q_r^s ($r = 1, 2, \ldots, R$) value. These problems are simultaneously solved by a dynamic programming procedure, with q^s referred to as a state variable.

First assume that VC^s and the elements of g'^s are constant. This implies that they are independent with respect to q^s and a^s. Accordingly $g'^s v^0$ and VC^s (denoted later as *FC,* fixed cost) can be dropped while maximizing z^*. Modifications of this assumption can be easily incorporated.

Assume that an activity fulfilling (3) has been found. Its addition to (1) and (2) (by the revised simplex method) increases the value of z in (1). Since (a) the coefficients q^s and a_i^s are discrete and finite and only a finite number of irrigation activities exists, and (b) the value of z increases in any LP-DP loop iteration, it follows that the process converges to optimality in a finite number of steps.

3. SUBSYSTEM II: THE SOIL MOISTURE SUBROUTINE

A necessary condition for determining an optimal irrigation policy is information about soil moisture variation over time and depth. Several methods to evaluate evapotranspiration and soil moisture fluctuation are available. They differ in sophistication and data requirements. A relatively simple model is sometimes adequate for prediction purposes.

The soil moisture-budgeting model applied here assumes that the evapotranspiration process can be expressed by the following relationship (Yaron et al., 1973; Shimshi et al., 1975):

$$(6) \qquad ET = -dw/dt = a + bw,$$

where $a < 0$, $b > 0$, w is the daily mean soil moisture content of the layer, and t is time in days. Soil moisture is expressed in terms of dry weight percentage [(weight of soil moisture/weight of dry soil) $\times$ 100]. ET is the daily evapotranspiration from the layer, equal to the reduction in w ($-dw/dt$), and a

and b are parameters. Note that for $w = PWP$ (permanent wilting point) $ET = 0$, and, therefore, $a = -b \cdot PWP$. Also, assume that the parameters of this function vary according to the "growth periods," as defined by the (phenological) development stages of the crop (Table 8.1). Accordingly, consider the following function:

$$(7) \qquad ET_j = -dw/dt = a_j + b_j w,$$

where $a_j < 0$, $b_j > 0$, and j is the index of growth period. Note that the growth period indirectly presents atmospheric evaporative conditions.

The parameters a_j and b_j are estimated by a computer search technique to achieve a good fit between the computed moisture values and the measured values as available. The subroutine can be calibrated for different locations and crops. It has been successfully applied to wheat (Yaron et al., 1973) and to sorghum (Bielorai and Yaron, 1978).

Table 8.1. Subdivision of the irrigation season

Days from Planting Date [a]	51-60	61-70	71-80	81-90	91-100	101-110	111-120	121-130
Peak season decade [b]	1	2	3	4	5	6	7	8
Month	May	June	June	June	July	July	July	Aug
Phenological stage of growth [c]	$j=1$	$j=1$	$j=2$	$j=2$	$j=3$	$j=3$	$j=4$	$j=4$
Evapotranspiration coefficients[d]								
a_j	-.4734	-.4734	-.5641	-.5641	-.6500	-.6500	-.7033	-.7033
b_j	.0278	.0278	.0332	.0332	.0382	.0382	.0414	.0414
Yield loss coefficients[d]								
d_j	0	0	0	0	4.05	4.05	1.81	1.82

[a] Planting date; 1 April; Harvest: October, November. No irrigation beyond 10 August
[b] Subperiods corresponding to restrictions in water availability
[c] The phenological stages being: 1) from planting to the beginning of flowering; 2) until the peak of flowering; 3) until the end of flowering; 4) until the beginning of bolls opening; 5-until the ripening of bolls (this stage is beyond the irrigation season)
[d] See text for explanation

4. SUBSYSTEM II: THE CROP RESPONSE SUBROUTINE

The general specification of the crop response function to soil moisture is (Bielorai and Yaron, 1978) $Y = f^*[SMI_j; AT_j]$, where j is index of growth period, SMI is soil moisture index, and AT is index of atmospheric conditions other than rainfall.

Several approaches are available for the actual specification of this function and for the choice of the independent variables. In this study, the soil moisture index in the root zone during the growing season was expressed by "stress days" or "critical days." A "critical day" was defined as one during which the soil moisture was depleted below a given percentage of available soil moisture (ASM) in the root zone. The number of "critical days," thus defined, was used as an explanatory variable in the response function.

The following actual response function was applied:

$$(8) \qquad Y = A - \Sigma d_j X_j,$$

where Y is crop yield (kg/ha); X_j, number of "critical days" in growth period j (with ASM below an empirically determined critical level); A, maximal yield obtainable for $X_j = 0$ for all j; d_j, loss of crop yield per critical day in growth period j (kg/ha).[1] (Further details are in Yaron et al., 1973; and Bielorai and Yaron, 1978.)

5. THE DYNAMIC PROGRAMMING MODEL

The DP model considers an activity unit as one hectare on which the irrigated crop under consideration is grown. The model takes given shadow prices of water and calculates (a) the optimal total quantity of water (q^s) to be allocated to one activity unit throughout the season, and (b) the optimal allocation over time of that water.

The planning horizon (irrigation season) extends over T days and is subdivided into J "decades" corresponding to subseasonal water supply restrictions. On each day, a decision regarding irrigation is considered.

The model has two discrete state variables: w_t^n, soil moisture level n (n = 1, ..., N) at the beginning of day t; and q_t^m, the mth quantity of irrigation water available for the one hectare plot under consideration (m = 1, ..., M). (Note that the initial possible values of q_t^m at $t = 1$ are equal to the number (R) of parametric values of q^s.) On each day, an irrigation decision, D_t^k, is considered, where D_t^k, is decision k (k = 1, ..., K) taken on day t, after the state of the plot $[w_t^n, q_t^m]$ has been observed. Note that $D_t^k \le q_t^m$.

The state of the hectare plot on day t $[w_t^n, q_t^m]$ and the decision D_t^k taken determine together (a) the state of the plot on the following day:

$$(9) \qquad w_{t+1}^{n*} = f_t\left(D_t^k \middle| \; w_t^n, \; q_t^m\right),$$

(10) $$q_{t+1}^{m*} = g_t\left(D_t^k \middle| \; q_t^m\right);$$

and (b) the "immediate loss" incurred $[h_t(D_t^k \mid w_t^k)]$. The immediate loss function $h_t^{k(\cdot)}$ contains the cost of irrigation, if applied, or the loss of ultimate yield sustained on day *t* if irrigation is not applied and soil moisture falls below the critical level.

The core of the DP model is the following recursive relationship:

(11) $$\Lambda_t\left(w_t^n, q_t^m\right) = \max_{D_t^k}\left[h_t\left(D_t^k \middle| w_t^n, q_t^m\right) + \Lambda_{t+1}\left(w_{t+1}^{n*}, q_{t+1}^{m*}\right)\right] ,$$

$t = 1, 2, \ldots, T - 1$, subject to (9) and (10), with $h_t^k(\cdot) \leq 0$; and

(12) $$\Lambda_T\left(w_T^n, \; q_T^m\right) = A \cdot P_y - FC$$

for all *n* and *m*, where P_y is the price per yield unit (net of harvest cost), and *FC* is the fixed cost per hectare excluding irrigation and harvest costs. Fixed cost (*FC*) can be ignored in the recursive maximization process.

The objective is to maximize the cumulative net income $\Lambda_1\left(w_1^n, \; q_1^m\right)$, for every *n* and *m* by applying a DP backward-induction procedure to (11), subject to (9) and (10) for $t = T - 1$, $T - 2$, and so on. Here $t = 1$ denotes the beginning of the growing season.

6. SPECIFICATION OF THE SYSTEM

This system has been applied to the analysis of the optimal allocation of water on a typical farm in the south of Israel (Ha'darom) during the peak irrigation season for cotton. Cotton is an important cash crop, and this problem is quite relevant to many Israeli farms. Preliminary empirical estimates of the cotton response to soil moisture for the region are available in Hebrew (Yaron and Dinar, 1978). The evapotranspiration coefficients (a_j and b_j) were synthesized from available data, and are plausible. The linear programming model is conventional. Only the water restrictions and cotton irrigation activities are shown in Table 8.2.

The growing season, its subdivision into water supply decades (previously defined), and the phenological stages of cotton growth are shown in Table 8.1. The phenological stages of growth are: $j = 1$ from planting to the beginning of flowering, $j = 2$ until the peak of flowering, $j = 3$ until the end of flowering, $j = 4$ until the beginning of boll opening, $j = 5$ until boll ripening. Table 8.1

also shows the correspondence between the decades, the yield loss coefficients (d_j) and the evapotranspiration coefficients (a_j , b_j).

For illustration, assume that all plots were planted on the same day (1 April). This assumption can be adapted easily to a realistic situation where the planting period extends over three to four weeks, and distinct plots can be differentiated according to planting dates. Soil moisture at planting was assumed to be at field capacity.

The basis for the assumed yield function was an empirical estimate derived from experimental data (Yaron and Dinar, 1978)

$$(13)\qquad Y = \underset{**}{5{,}670} - \underset{*}{40.5X_3} - \underset{*}{18.2X_4}\,,$$

where R^2 = .61, Y is yield of raw cotton [kg/ha]; X_3, X_4 is number of critical days (soil moisture below 40 percent of available soil moisture in the root zone) during growth periods 3 and 4; and **, * are significance at 1 percent and 5 percent probability level, respectively.

The moisture state variable, w_t^n, was expressed in terms of discrete levels ranging from FC (field capacity) to PWP (permanent wilting point). For the second state variable, q_t^m (quantity of water available for the irrigation of the particular activity plot on day t), 11 discrete levels were distinguished ranging from 0 to 4,000 m^3 per hectare with increments of 400 m^3/ha between states. Altogether, 1,221 states were distinguished ($11 \times 111 = 1{,}221$).

Table 8.2. Water constraints and initial cotton irrigation activities

Decade	Basic Quantity	Options to Buy		Activity No.			Shadow Price at First LP Optimal Solution
		Quantity	Price	1	2	3	
	(000 m^3)	(000 m^3)	(IL/m^3)[a]		(m^3/ha)		(IL/m^3)[a]
1. May 21-31	80	20	2.2	700		500	42.7
2. June 1-10	80	20	2.5		800		
3. June 11-20	80	20	2.5	700			
4. June 21-30	80	20	3.6			700	2.3
5. July 1-10	100	20	3.6		1000		21.9
6. July 11-20	110	20	4.0	1200			
7. July 21-31	110	20	4.0		1000		0.6
8. Aug 1-10	100	20	3.0	1000			
Seasonal Total	700[b]			3600	2800	1200	3.4
Activity level at first LP optimal solution (ha)				31	102	116	

[a]At 1979 price level, one IL (Israeli Lira) = 3 US cents, approximately.
[b]Note that the seasonal total is not the sum of the decade constraints. There is some flexibility among the decades.

The decision variables D_t^k were defined in discrete levels of water inputs on day *t*, $D_t^k = 0, 400, 800, 1{,}200\ m^3/ha$.

The soil moisture transformation function was

$$(14) \qquad W_{t+1}^{n*} = w_t^n - \left(a_j + b_j w_t^n\right) + \gamma D_t^k,$$

where γ is a coefficient transforming water quantities per hectare (D_t^k) into soil moisture percentages in the root zone. Note that γ is an empirical parameter depending on soil properties and depth. The values of a_j and b_j ($j = 1, 2, \ldots, 4$) are presented in Table 8.1.

The relationship between q_t^m and q_{t+1}^{m*} is

$$(15) \qquad q_{t+1}^{m*} = q_t^m - D_t^k.$$

The immediate loss function contains two elements: (a) Cost of irrigation, if applied, ($D_t^k \cdot PW_t$). Note that the water price on day *t*, may vary from one "decade" to another. After the first run of the LP problem, PW_t values are given by the shadow prices of water. (b) The loss of yield value, if incurred, is $LR_t = \delta_t \cdot d_j \cdot P_y$, where $\delta_t = 1$ if the soil moisture on day *t* is less than or equal to a critical level (40 percent of the available moisture in the root zone), and $\delta_t = 0$, otherwise.

Accordingly the immediate loss function is

$$h_t^k(\cdot) = -\left(D_t^k \cdot PW_t + \delta_t \cdot d_j \cdot P_y\right).$$

7. EMPIRICAL ANALYSIS

The typical farm in the south region of Israel (Ha'darom) grows irrigated fruit crops, and cotton. The irrigable land residual is allocated to unirrigated wheat. The water supply constraints are shown in the two left columns of Table 8.2. These constraints represent the total water quantity and decade supply quantities throughout the season. The cost of one cubic meter is 2.00 Israeli Lira.[2] The farm can also purchase limited additional amounts of water (20,000 m^3 per decade) at an additional cost, ranging between IL 2.20 and IL 4.00 per m^3, depending on the decade. The farm's other constraints are land and the fixed areas of the fruit crops.

At first, three cotton irrigation-scheduling activities were formulated (Table 8.2) and incorporated into the first LP problem. The optimal solution levels of these activities and the shadow prices of water are shown in the bottom line and the right-hand column, respectively, of Table 8.2. Considerable variation in the shadow prices of water at the various decades is evident.

The shadow prices of water from this LP solution were substituted into the DP problem to begin the LP-DP loop. Results of the consecutive loop computations are shown in Tables 8.3 and 8.4.

Table 8.3 shows the initial cotton activities numbered 1-3 (loop No.1). The new activities 4-7 were sequentially generated by the DP procedure in response to the shadow prices of water from the LP solution in the previous loop. Table 8.4 shows cotton irrigation, the total cotton area, and farm income at the sequential LP solutions. Note that the final optimal program (loop No. 5) includes 48 hectares of unirrigated cotton.

These results indicate that (a) the farm's income rose by approximately 11 percent in transition from loop No. 1 to loop No. 5, while the income derived from cotton, only, rose by 19 percent; (b) incorporating new cotton activities into the farm's plan "balances" water use and smoothes out differences among the shadow prices of water over the various decades; (c) the total irrigated cotton area rose from 248 ha in loop No. 1 to 262 hectares in the final solution, while the unirrigated cotton area changed from 61 hectares in loop No. 1 to 48 ha in loop No. 5; and (d) the optimal irrigation intensities in the final program range from 1200 m^3/ha up to 2800 m^3/ha.

8. UPDATING THE IRRIGATION PROGRAM DURING THE SEASON

This approach assumes certainty about weather conditions. This assumption is valid for rainfall, which, under Mediterranean conditions, is confined to the winter season before cotton is planted. Uncertainty about evaporative conditions can be incorporated into the model but at a considerable cost. Instead, midseason adjustments are suggested. The need for midseason adjustments also may arise because of disturbances in executing the irrigation schedule. For example, several weeks after cotton planting, the farm might be informed that its peak season water supply must be reduced because of unexpected difficulties in the water supply system.

Table 8.3. Generation of new irrigation activities and shadow prices of water of the LP-DP consecutive loop

Loop No.			(1)		(2)		(3)		(4)		(5)	
Activity	1	2	3		4		5		6		7	
Decade		Water Input (m^3/ha)		Shadow Price (IL/m^3)	Water Input (m^3/ha)	Shadow Price (IL/m^3)	Water Input (m^3/ha)	Shadow Price (IL/m^3)	Water Input (m^3/ha)	Shadow Price (IL/m^3)	Water Input (m^3/ha)	Shadow Price (IL/m^3)
1. May 21-31	700		500	42.7		29.7		13.9	800	16.7		15.2
2. June 1-10		800			800	29.7	400	19.7		16.7	400	15.2
3. June 11-20	700				800	14.9	400	16.3	800	16.7	400	16.9
4. June 21-30			700	2.3		12.2	400	19.7		15.2	400	15.5
5. July 1-10		1,000		21.9			400				400	
6. July 11-20	1,200				400				400			
7. July 21-20		1,000		0.6	400		400		400		400	
8. August 1-10	1,000											
Total per season	3,600	2,800	1,200	3.4	2,400	3.0	2,800[a]	5.8	2,400	6.7	2,000	7.1

[a]Including 800 m^3 before the peak season

Two types of midseason adjustments can be identified: intrafarm and interfarm. Intrafarm midseason adjustments in water allocation among various plots of cotton and other crops can be considered. Recall that the farm under consideration grows cotton, fruit crops, and unirrigated wheat. The approach assumes a fixed schedule of irrigation of fruit crops aimed at achieving the maximal yield (rather than a yield level at which the marginal value product of water equals its alternative cost). This is so because of the high income for fruit crops and the chance for error in decisions regarding their irrigation schedule. In case of an error leading to overirrigation, the economic loss (alternative cost of water) will be small in comparison with the income derived. On the other hand, in case of an error leading to underirrigation, the value of the sustained yield will be high relative to the alternative cost of water saved. Accordingly, the irrigation schedule of fruit crops remains fixed and the adjustment analysis is restricted to cotton only.

For illustration, we present a hypothetical "adjustment analysis" for the farm under consideration. First, an "adjustment" program is solved before the beginning of the peak season (21 May). All expenses related to cotton before that date are considered "sunk." At this point, they are equal for all cotton plots and amount to 49,000 Israeli pounds per hectare. The water supply to the farm in the fourth decade (21-30 June) must be reduced by 20,000 m^3 (20 percent of the decade's quota). The cotton growth path, the irrigation schedule, and the evaporative conditions, until 21 May, were as anticipated in the preplanting program. The irrigated total potential irrigated cotton area is given and cannot exceed 262 hectares, but it can be reduced. The results of the analysis (not detailed here) suggest irrigation schedule adjustments for cotton including a necessary water use reduction in the fourth decade. The shadow prices of water change generally upward in comparison to those derived from the preplanting program.

Another need for a midseason adjustment program may result from a drastic change in expected average evaporative conditions. For example, a sequence of three unusually hot and dry days may cause unexpected soil moisture depletion of about 24,000 cubic meters.[3]

The need for midseason adjustments emphasizes the importance of a flexible computer program for making quick computations. Consider interfarm adjustments. The shadow prices of water, several times higher than its actual cost to farmers, generate demands for additional water above the institutional quotas for the farm. This is exacerbated by possible unexpected midseason difficulties in executing the original irrigation plan.

Table 8.4. Hectares of cotton at consecutive LP solutions

	Activity Number[a]								
Loop No.	1	2	3	4	5	6	7	Cotton Area (Total ha.)	Farm Income 10^6 IL
	----------------------(ha.)------------------								
(1)	30	102	116	13.34				248	12.39
(2)	30	26	117	13.47				249	13.34
(3)	40	35	103	13.68	25			258	13.47
(4)	0	46	85	13.73	57	46		262	13.68
(5)	0	46	85	28	0	46	57	262	13.73

[a]See Table 8.3 for the details of water inputs of these activities

Economists traditionally advocate using a price mechanism for interfarm water allocation systems. This is sensible because of obvious inefficiencies of prevailing institutional allocation schemes. However, the above analysis leads us to question water allocation by prices for short-run midseason adjustments.

Consider a region with interfarm water allocation by competitive bidding. Assume also that the share of cotton in the total demand for water is significant. Cotton-growing farmers adjust their irrigation schedules during the season, and their demand prices for water vary accordingly. Under these conditions uncertainty with respect to water prices will be generated. Such uncertainty may have a negative effect on producers' welfare. This hypothesis is based on an analogy by Oi (1961), Sandmo (1971) and others who have found that under certain conditions product price instability reduces producer welfare. Further study is needed for the exploration of the above hypothesis and analysis of the effect of short and long-term water price instability on farmers' welfare.

9. CONCLUSION

The paper presents a system analysis approach to the allocation of scarce water during peak seasons to alternative crops and plots using soil-moisture response functions for the key crops. This approach provides an irrigation-scheduling program for the farms during the peak season, taking into account overall farm restrictions and the shadow prices of water and other resources.

The results provide a schedule for an optimal allocation of water among crops and over time. The sample farm analysis suggests that there is considerable potential for improving irrigation management and for increasing income.

The approach assumes certainty about weather and other relevant variables. Uncertainty can be incorporated into the model at the cost of considerably expanding it. Instead, follow-up, mid-season adjustment analyses are suggested, and their economic implications are discussed briefly.

In order to apply this approach to extension use[4], two modifications are needed: (a) the soil moisture and crop response subroutines should be calibrated and adapted to any particular set of local conditions, and (b) restrictions emerging from the hydraulics of the farm's water supply system should be incorporated. A prerequisite for application of the approach is an easy access to computer facilities, either directly or through extension service channels.

REFERENCES

Bielorai, H. and D. Yaron. Methodology and empirical estimates of the response function of sorghum to irrigation and soil moisture. *Water Resources Bulletin,* 14:966-77. 1978.

Blank, H. G. Optimal irrigation decisions with limited water (PhD Thesis). Colorado State University. Ft. Collins, CO. 1975.

Dantzig, G. B. "Linear Programming and Extensions." Princeton, NJ. Princeton University Press. 1963.

Dinar, A. Use of interdisciplinary models to extension. Agricultural Administration and Extension, 24(3):165-176. 1987.

Dudley, N. I., D. T. Howell and W. F. Musgrave. Irrigation planning 2: choosing optimal acreages within a season. *Water Resources Research,* 7:1051-63. 1971.

Hall, W. A. and W, S. Butcher. Optimal timing of irrigation. *American Society of Civil Engineers Journal of Irrigation and Drainage Division, IR2* 92:267-275. 1968.

Howell, T. A., E. A. Hiler and D. L, Reddell. Opitimizing water use efficiency under high frequency irrigation—II system simulation and dynamic programming. *American Society of Agricultural Engineers,* 18:879-887. 1978.

Oi, N. Y. The desirability of price instability under perfect competition. *Econometrica,* 29:59-64. 1961

Sandmo, A. On the theory of the competitive firm under price Uncertainty. *Agricultural Economic Research,* 61:65-73. 1971

Shimshi, D., D. Yaron, E. Bresler, M. Weisbrod, and G. Strateener. Simulation model for evapotranispiration of wheat: empirical approach. *American Society of Civil Engineers Journal of Irrigation and Drainage Division,* IR1, 101:1-12. 1975

Yaron, D. Estimation and use of water production functions in crops. *American Society of Civil Engineers Journal of Irrigation and Drainage Division,* IR2 97:291-303. 1971.

Yaron, D. and A. Dinar. Optimal timing of cotton irrigation on a daily basis. Center for agricultural economics research report. The Hebrew University of Jerusalem, Rehovot, Israel. 1978. (Hebrew).

Yaron, D., E. Bresler, H. Bielorai and B. Harpinist. A model for optimal irrigation scheduling with saline water. *Water Resources Research,* 16:257-262. 1980.

Yaron, D., G. Strateener, D. Shimshi, and M. Welsbrod. Wheat response to soil moisture and the optimal irrigation policy under conditions of unstable rainfall. *Water Resources Reearch,* 9:1145-1154. 1973.

---

[1] An alternative formulation is the exponential function $Y = A\prod_j b_j^{x_j}$ with $b_j (0 < b_j \leq 1)$ being a coefficient of crop reduction in growth period j with the other variables and parameters as above.

[2] At 1979 price level, one IL = 3 US cents, approximately.

[3] Evapotranspiration higher than expected by 3 mm per day is equivalent to 30 m^3 per hectare per day and to 23,580 m^3 per 262 hectares per three days.

[4] See application in Dinar (1987).

9

THE VALUE OF INFORMATION ON THE RESPONSE FUNCTION OF CROPS TO SOIL SALINITY*

Eli Feinerman
University of California, Berkeley, California, USA

Dan Yaron
The Hebrew University of Jerusalem, Rehovot, Israel

1. INTRODUCTION

Agricultural production involves a large number of random variables, many of which are physical and biological functions connected with the production process. The knowledge of the biological response function of crop yield to soil salinity is essential in decision-making regarding irrigation with saline water. In this paper we investigate (analytically and empirically) the expected profitability to farmers (the decision makers) of acquiring additional information on this biological function. The true values of the parameters of the response function are usually unknown to the decision maker, and therefore he uses the estimates of the parameters and may become a victim of a suboptimal solution. The deviation from the optimum may be measured by a loss function and the calculation of its expectation. The estimates of the parameters (which are arguments in the loss function) are based on *a priori* information available to the decision maker. He can acquire additional information that will reduce the variances of these estimates and, hence, will improve his ability to choose a suitable strategy with resulting decrease of the expected loss (or, equivalently, increase of the expected profit). Expected value of sample information (EVSI) is defined as the difference between the reduction of the expected

*Permission to publish this chapter was granted by Academic Press, Inc. The chapter was originally published under a similar title in *The Journal of Environmental Economics and Management,* 10:72-85, 1983.

value of the loss function due to the additional information and the cost of its acquisition. The optimal number of observations to be acquired is the one that maximizes EVSI.

An accepted hypothesis among soil researchers states that the yield of a given crop is a function of, among other variables, the average soil salinity in the root zone during the growing season. *Ceteris paribus,* increase of the average soil-salinity level slows down the rate of growth and reduces crop yield (e.g. Bernstein, 1980; Maas and Hoffman, 1977; Shalhevet and Bernstein, 1968). The relationships between the soil-salinity level and the reduction of crop yields has been dealt with previously. Some researchers have shown these relationships in the form of tables (Bernstein, 1980; Bierhuizen, 1969; DeForges, 1970); in other studies the response function was hand-fitted to the available observations (Polovin, 1974; Yaron and Olian, 1973). Few publications report estimates of continuous response functions based on the "best linear unbiased estimates" (BLUE) criterion (Nouri, 1970; Shalhevet and Bernstein, 1968; Yaron et al., 1972).

A detailed discussion of the salinity response function is found in Maas and Hoffman's article (1977). Using data on relative yield losses due to salinity with respect to a wide range of crops, that is, fruit crops, field crops, and vegetables, they hypothesized a threshold soil salinity level, beyond which a linear decrease in relative yield is obtained. The critical threshold hypothesis is also presented by Bernstein (1980).

A broad theoretical presentation of decision theory, value of information, and the Bayesian approach can be found in DeGroot (1970) and Pratt et al. (1965). A number of studies deal with the value of information in farm management (Maddock, 1973; Ryan and Perrin, 1974) as well as in the management of water resources (Davis and Dvoranchik, 1971; Klemes, 1977). It should be pointed out that most of these articles did not deal explicitly with the choice of the optimal estimate or with the optimal size of the sample. Furthermore, the articles that dealt with the management of irrigation systems did not refer to water quality.

To calculate the expected profitability to farmers, of the additional acquired knowledge about the biological response function to soil salinity, a three-step procedure was followed: a) A switching regression approach (e.g., Quandt, 1960) was used to estimate the parameters of the response function model, according to the response function as formulated by Maas and Hoffman; b) An optimization irrigation model for a monoculture farm was developed, aimed at determining the optimal quantity of irrigation water from a given source for soil leaching (to reduce salinity); c) A loss function was defined, the EVSI was calculated, and the optimal number of additional needed observations was determined. At each stage, an empirical analysis using data from potato field experiments carried out by Sadan and Berglas (1980) in the Negev area of Israel is presented.

2. ESTIMATION OF THE RESPONSE FUNCTION

The following model was formulated (see Figure 9.1):

$$\begin{aligned} Y &= b_0 + U_1 && \text{if} \quad S \leq S_0 \\ &= b_1 + aS + U_2 && \text{if} \quad S > S_0 \end{aligned}$$

subject to

(1) $b_0 = aS_0 + b_1$

where S is the average soil-salinity level in the root zone [meq Cl/l] during the growing season; S_0 the threshold salinity of the soil [meq Cl/l]; Y the yield in tons per hectare (ha); U_1, U_2 the independent random variables, normally distributed with zero expectation; and b_0, b_1, a, S_0, the (unknown) parameters of the response function satisfying (1).

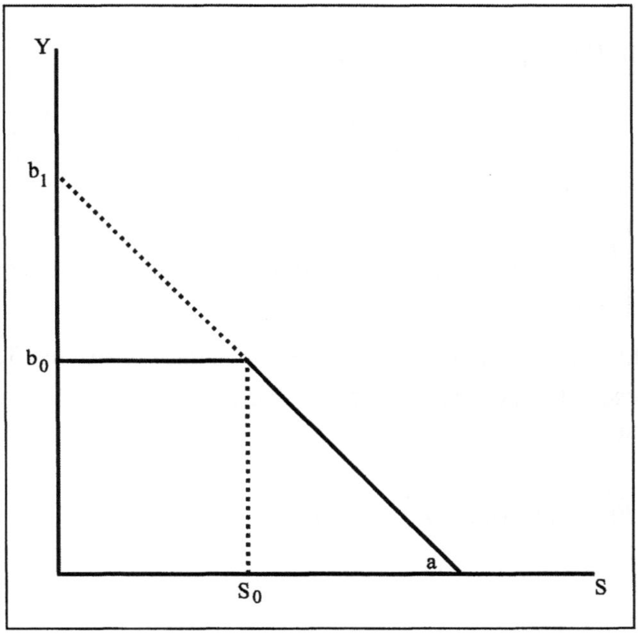

Figure 9.1. The response function

Assume that we have T observations (S_i, Y_i) for estimating the parameters. Arranging the S_i in increasing order, we have

$$S_1 \le S_2 \cdots \le S_t \le \cdots \le S_T$$

where

$$\begin{matrix} S_i \le S_0 \\ S_i > S_0 \end{matrix} \quad \text{for} \quad \begin{matrix} i \le t \\ i > t \end{matrix} \qquad (t \text{ unknown})$$

A regression model can be formulated:

$$(2) \qquad \begin{aligned} Y_i &= b_1 + aS_0 + U_{1i} \quad &\text{if} \quad & S_i \le S_0; \quad i \le t \\ &= b_1 + aS_i + U_{2i} \quad &\text{if} \quad & S_i > S_0; \quad i > t\,. \end{aligned}$$

This model assumes independent normally distributed random deviations with mean zero and $T \times T$ diagonal variance-covariance matrix Ω. Its first t diagonal elements are $V(U_{1i}) = \sigma_1^2$ and the other $(T-t)$ elements are $V(U_{2i}) = \sigma_2^2$

The logarithm of the likelihood function $L(Y|\ S_0,\ t)$, given S_0 and t, is

$$(3) \qquad \ln\ L(Y|\ S_0, t) = -T\ln\sqrt{2\pi} - t\ln\sigma_1 - (T-t)\ln\sigma_2 - \frac{1}{2\sigma_1^2}\sum_{i=1}^{t}(Y_i - aS_0 - b_1)^2 - \frac{1}{2\sigma_2^2}\sum_{j=t+1}^{T}(Y_j - aS_j - b_1)^2$$

Let $\hat{\beta} = \begin{bmatrix} \hat{b}_1 \\ \hat{a} \end{bmatrix}$ be the vector of maximum likelihood estimates (MLE), that is, $\hat{b}_1$, $\hat{a}$ which maximizes (3). Because the square terms in (3) have negative signs, these estimates are identical to the least-square estimates (given S_0, t), so that one can write

$$(4) \qquad \hat{\beta}(S_0,\ t) = (Z'\hat{\Omega}^{-1}Z)^{-1} Z\hat{\Omega}^{-1}Y$$

where

$$Z = \begin{bmatrix} 1 & S_0 \\ \vdots & \vdots \\ 1 & S_0 \\ 1 & S_{t+1} \\ \vdots & \vdots \\ 1 & S_T \end{bmatrix}; \qquad Y = \begin{bmatrix} Y_1 \\ \vdots \\ Y_t \\ Y_{t+1} \\ \vdots \\ Y_T \end{bmatrix}$$

(For a more detailed discussion see Feinerman, 1980).

The MLE of σ_1^2 and σ_2^2 can be obtained by differentiating (3) and equating to zero:

$$(5) \qquad \hat{\sigma}_1^2(S_0,\ t) = \sum_{i=1}^{t} \left(Y_i - \hat{a}S_0 - \hat{b}_1\right)^2 / t$$

$$(6) \qquad \hat{\sigma}_2^2(S_0,\ t) = \sum_{j=t+1}^{T} \left(Y_j - \hat{a}S_j - \hat{b}_1\right)^2 / (T-t).$$

Equations (4)-(6) form a set of four equations in four variables; these can be solved numerically with the aid of a computer (Harville, 1977).

Substituting these estimates into (3) yields

$$(7) \qquad \ln \hat{L}\left(Y \middle|\ S_0,\ t\right) = -T \ln\sqrt{2\pi} - t \ln \hat{\sigma}_1 - (T-t)\ln\hat{\sigma}_2 - \frac{1}{2}T.$$

Finally the estimates of S_0, t can be obtained as follows:

Step 1. Between every two consecutive observations S_{i-1} and S_i (starting at the third observation and stopping three observations before the end),[1] (7) is maximized numerically over S_0 as

$$i = 3,\ \ldots,\ T-3 \qquad L_i\left(S_{i-1},\ S_i\right) = \max_{S_{i-1} < S_0 \le S_i} \left\{\ln \hat{L}\left(Y \middle|\ S_0,\ i\right)\right\}$$

$$L_{T-2}\left(S_{T-3},\ S_{T-2}\right) = \max_{S_{T-3} < S_0 < S_{T-2}} \left\{\ln \hat{L}\left(Y \middle|\ S_0,\ T-2\right)\right\}$$

Step 2. The optimal estimates of S_0, t are $\hat{S}_0$, $\hat{t}$ which satisfy

$$\ln \hat{L}\left(Y \middle|\ \hat{S}_0,\ \hat{t}\right) = L_t\left(S_{\hat{t}-1},\ S_{\hat{t}}\right) = \max_{2<i<T-2} L_i\left(S_{i-1},\ S_i\right).$$

Let $\hat{\theta} = \left[\hat{S}_0,\ \hat{b}_1,\ \hat{a}\right]$. From the properties of MLE (e.g., Maddala, 1977), under fairly general conditions, $\hat{\theta}$ is asymptotically normally distributed with mean $\theta = \left[S_0,\ b_1,\ a\right]$ and variance-covariance matrix

$$\Sigma_{\hat{\theta}} = \left[E\left[\frac{-\partial^2 \ln L}{\partial \theta_i \partial \theta_j} \right]^{-1} \right].$$

The variances and covariances can be calculated as functions of the observations S_i and of a, S_0, t, T, σ_1^2, σ_2^2.

2.1. Empirical Results

The response-function estimates are based on the experimental results of Sadan and Berglas (1980). Their one-year experiment, conducted in the Northern Negev of Israel, provided a total of 17 observations (Y_i, S_i) which are presented in Table 9.1.

Table 9.1. Relationships between average soil salinity (S_i) and potato yield (Y_i)—the Negev area

	1	2	3	4	5	6	7	8	9
Y_i^a	45.4	46.2	46.2	48.7	45.8	48.1	33.2	45.4	38.9
S_i^b	4.015	4.234	4.964	6.058	7.664	8.394	12.263	12.401	12.774
	10	11	12	13	14	15	16	17	
Y_i	31.8	34.3	33.6	34.9	25.3	28.5	29.1	32.4	
S_i	12.847	16.642	18.102	18.832	19.781	20.292	22.263	22.701	

[a] Yield [tons/ha].
[b] Average soild-salinity level [meq Cl/l] in the root zone (0-60 cm), during the growing period.

The following estimates for the response-function parameters (2) were obtained by substituting the data in Equations (4)-(6):

	Asymptotic standard deviation
$\hat{S}_0 = 6.054$ [meq Cl/l]	(1.92)
$\hat{b}_1 = 52.55$ [tons/ha]	(3.2)
$\hat{a} = -1.09$ [tons/ha]/[meq Cl/l]	(0.2)

Accordingly, $\hat{b}_0 = 45.95$ [tons/ha] and the estimated response function is

$$i = 1, \ldots, 17 \qquad \hat{Y}_i = 45.95 \quad \text{if} \quad S_i \le 6.054$$
$$= 52.55 - 1.09 S_i \quad \text{if} \quad S_i > 6.054$$

with other relevant statistics being

$$\hat{t} = 3, \quad \sigma_1^2 = 0.14 \text{ [tons/ha]}^2; \; \hat{\sigma}_2^2 = 15.78 \text{ [tons/ha]}^2; \; R^2 = 0.78 .$$

The scatter diagram of the observations and the fitted regression line are shown in Figure 9.2. It should be noted that this response function (2) was compared with estimations of alternative formulations (all of them monotonically decreasing, continuous, and differentiable functions) and was judged "best" (Feinerman, 1980) using "adjusted *R*-square" criterion.

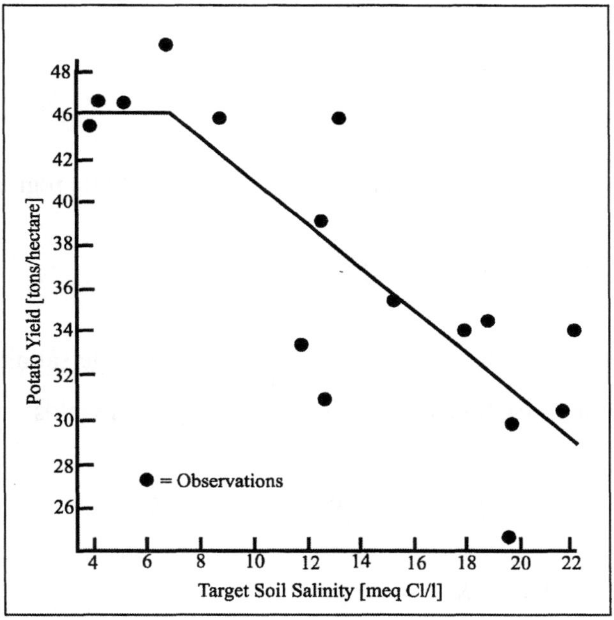

Figure 9.2. The estimated response function of potatoes

3. OPTIMAL USE OF IRRIGATION WATER FROM A GIVEN SOURCE FOR A GIVEN CROP (POTATOES)

Consider the yield expectation

$$\text{(8)} \qquad \begin{aligned} EY &= aS_0 + b_1 && \text{if} \quad S \leq S_0 \\ &= aS + b_1 && \text{if} \quad S > S_0 \end{aligned}$$

Let $\bar{S}$ be the present soil-salinity level, $\bar{S} > S_0$. Assume that there is a source of good quality water of salinity Cl [meq Cl/l]. Leaching the soil with Q m^3 of this water per hectare will reduce the salinity to $\underline{S}$, where (see Bresler, 1967)

$$\underline{S} = \frac{QCl}{(N + (\gamma/2)Q)} + \bar{S}\frac{(N - (\gamma/2)Q)}{(N + (\gamma/2)Q)}$$

(N and γ are known soil parameters).

Isolating Q we obtain

$$(9) \qquad Q(\underline{S}) = \frac{N(\bar{S} - \underline{S})}{(\gamma/2)(\bar{S} + \underline{S}) - Cl}$$

A good empirical approximation is obtained by the quadratic regression

$$(10) \qquad Q(\underline{S}) = K_1 + K_2^*(\bar{S} - \underline{S}) + K_3^*(\bar{S} - \underline{S})^2.$$

Let $S = (\bar{S} + \underline{S})/2$ be an approximation of the average salinity level (before and after leaching). Substituting $\underline{S} = 2S - \bar{S}$ in (10) yields

$$Q(S) = K_1 + K_2(\bar{S} - S) + K_3(\bar{S} - S)^2.$$

where $K_2 = 2K_2^*$, $K_3 = 4K_3^*$.

Let P be the cost of the water supply in dollars per cubic meter (m^3). The cost of leaching the soil (with Q m^3/ha) denoted by $C(\bar{S},\ S)$ is

$$(11) \qquad C(\bar{S},\ S) = PQ(S) = PK_1 + PK_2(\bar{S} - S) + PK_3(\bar{S} - S)^2.$$

A profit function is defined as

$$(12) \qquad \begin{aligned} \pi &= R_1(aS_0 + b_1) - R_2 - C(\bar{S},\ S) && \text{if} \quad S \le S_0 \\ &= R_1(aS + b_1) - R_2 - C(\bar{S},\ S) && \text{if} \quad S > S_0 \end{aligned}$$

where R_1 is the net income in dollars per unit of yield (ton) as a function of the yield (revenue, less variable cost dependent on yield, such as harvesting, grading, packing, and transportation); and R_2 is the variable costs per hectare, independent of yield. By substituting (11) into (12) and then equating the derivative $\partial\pi / \partial S$ (for $S > S_0$) to zero, S^* which maximizes (12) is obtained:

$$S^* = \overline{S} + \frac{R_1 a + PK_2}{2PK_3}.$$

Since it is obvious that $S^* \geq S_0$, it can be written

$$(13) \qquad S^* = \max\left(S_0,\ \overline{S} + \frac{R_1 a + PK_2}{2PK_3} \right).$$

And, by substituting MLE $\hat{a},\ \hat{S}_0$ for the unknown parameters, $a,\ S_0$

$$(14) \qquad \hat{S}^* = \max\left(\hat{S}_0,\ \overline{S} + \frac{R_1 \hat{a} + PK_2}{2PK_3} \right).$$

3.1. Empirical Results

The empirical approximation (10) to the leaching function (9), was achieved by dividing the relevant range of soil salinity into a large number of discrete points, calculating the value of $Q(S)$ by (9) for each point and estimating the regression line (10).

For $\overline{S} = 20$ [meq Cl/l] (a somewhat high initial soil-salinity level was chosen to emphasize the need of soil leaching), $N = 3500$ [m^3/ha], $\gamma = 0.7$ (average irrigated-soil parameters in the study area), and Cl = 5 [meq Cl/l], the following estimates were obtained

$$K_1 = 132, \qquad K_2 = 526, \qquad K_3 = 146, \qquad K_2^* = 236, \qquad K_3^* = 36.5.$$

Equations (9) and (10) are presented in Figure 9.3 on the same set of axes ((9) is marked by the numeral 1, and (10) by the numeral 2). With $P = 0.1$ [US\$/m^3]^2; $R_1 = 161$ [US\$/ton], the following values for potatoes were obtained:

$$\hat{S}^* = 15.8 \text{ [meq Cl/l]} \quad Q(\hat{S}^*) = 4917 \text{ [m}^3\text{/ha]}.$$

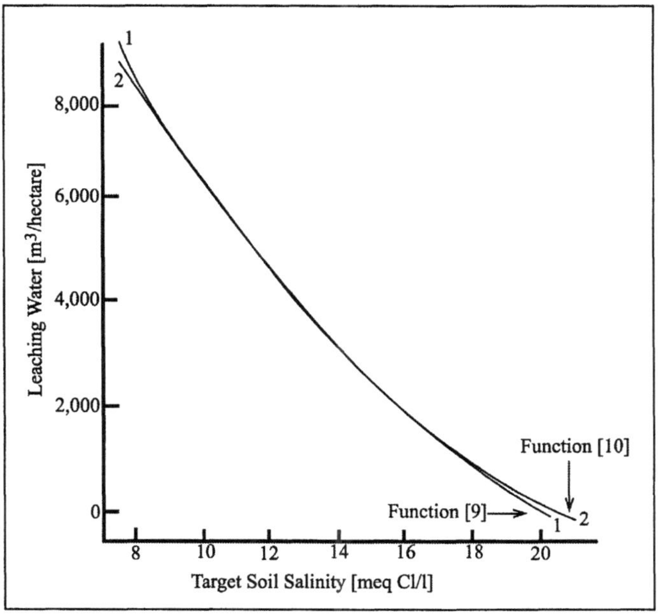

Figure 9.3. Quantity of leaching water (*Q*) as a function of the target soil salinity (*S*) (initial soil salinity $\overline{S} = 20$ meq Cl/l)

4. THE LOSS FUNCTION AND THE VALUE OF ADDITIONAL INFORMATION

In the following section, the loss function and its possible situations are defined, the EVSI is calculated and the optimal sample size is determined (taking into consideration the desired spread of the additional observations).

Generally speaking, the calculation of $Q(\hat{S}^*)$ is a decision problem in which the decision made by the (risk neutral) decision maker is based on his estimate of the value of some unknown parameters. The loss that he incurs will reflect the discrepancy between the true value and his estimate. For this reason, the loss function of the profit maximizer farmer is assumed to have the form[3]

$$\text{(15)} \qquad \text{LOSS}\left(\theta,\ \hat{\theta}\right) = \pi(\theta / \theta) - \pi\left(\hat{\theta} / \theta\right)$$

where $\pi\left(\hat{\theta} / \theta\right)$ is the value of the profit function when the decision making is based on the parameters estimates $\left(\hat{\theta}\right)$ being used in the model with true parameters θ. The profit function has a value of $\pi(\theta / \theta)$ when the true values of the parameters are used.

One may distinguish between eight alternatives associated with the possible values of the loss function based on all possible combinations of the relationships between $\pi(\hat{\theta}/\theta)$ and S_0, $\hat{S}^*$ and $\hat{S}_0$, S^* and S_0. But, four of them can be disregarded, since $\hat{a}$, $\hat{S}_0$ are consistent estimates (being MLE) and therefore tend to equal a and S_0 respectively, so

(16) $$P_r\left(S^* > S_0 \text{ and } \hat{S}^* \le \hat{S}_0\right) \rightarrow 0$$

(17) $$P_r\left(S^* \le S_0 \text{ and } \hat{S}^* > \hat{S}_0\right) \rightarrow 0.$$

The four remaining alternatives can be described as follows.

(a) $\hat{S}_0 > S_0$, $S^* > S_0$, $\hat{S}^* > \hat{S}_0$. Substituting in (15) yields

$$\begin{aligned}
\text{LOSS}\left(\theta,\ \hat{\theta}\right) = & \left\{ R_1 a\left(\bar{S} + \frac{R_1 a + PK_2}{2PK_3}\right) + R_1 b_1 - R_2 \right. \\
& \left. - P\left[K_1 + K_2\left(\bar{S} - \frac{R_1 a + PK_2}{2PK_3} - \bar{S}\right) + K_3\left(\bar{S} - \frac{R_1 a + PK_2}{2PK_3} - \bar{S}\right)^2\right]\right\} \\
& - \left\{ R_1 a\left(\bar{S} + \frac{R_1 \hat{a} + PK_2}{2PK_3}\right) + R_1 b_1 - R_2 \right. \\
& \left. - P\left[K_1 + K_2\left(\bar{S} - \frac{R_1 \hat{a} + PK_2}{2PK_3} - \bar{S}\right) + K_3\left(\bar{S} - \frac{R_1 \hat{a} + PK_2}{2PK_3} - \bar{S}\right)^2\right]\right\} \\
= & \frac{R_1^2 (\hat{a} - a)^2}{4PK_3}.
\end{aligned}$$

Similarly[4]

(b) $\hat{S}_0 \le S_0$, $S^* > S_0$, $\hat{S}^* > \hat{S}_0$

$$\text{LOSS}\left(\theta,\ \hat{\theta}\right) = \frac{R_1^2 (\hat{a} - a)^2}{4PK_3}.$$

(c) $\hat{S}_0 \le S_0$, $S^* = S_0$, $\hat{S}^* = \hat{S}_0$

$$\text{LOSS}\left(\theta,\ \hat{\theta}\right) = \left\{ R_1 \left(aS_0 + b_1\right) - R_2 - P\left[K_1 + K_2\left(\bar{S} - S_0\right) + K_3\left(\bar{S} - S_0\right)^2\right]\right\}$$

$$-\left\{R_1\left(aS_0+b_1\right)-R_2-P\left[K_1+K_2\left(\overline{S}-\hat{S}_0\right)+K_3\left(\overline{S}-\hat{S}_0\right)^2\right]\right\}$$

$$=-PK_2\left(\hat{S}_0-S_0\right)-PK_3\left[\left(\overline{S}-S_0\right)^2-\left(\overline{S}-\hat{S}_0\right)^2\right].$$

(d) $\hat{S}_0 > S_0,\ S^* = S_0,\ \hat{S}^* = \hat{S}_0$

$$\text{LOSS}\left(\theta,\ \hat{\theta}\right)=\left\{R_1\left(aS_0+b_1\right)-R_2-P\left[K_1+K_2\left(\overline{S}-S_0\right)+K_3\left(S-S_0\right)^2\right]\right\}$$

$$-\left\{R_1\left(a\hat{S}_0+b_1\right)-R_2-P\left[K_1+K_2\left(\overline{S}-\hat{S}_0\right)+K_3\left(\overline{S}-\hat{S}_0\right)^2\right]\right\}$$

$$=-\left(R_1a+PK_2\right)\left(\hat{S}_0-S_0\right)-PK_3\left[\left(\overline{S}-S_0\right)^2-\left(\overline{S}-\hat{S}_0\right)^2\right].$$

Using indicator functions the loss function is written concisely as

(18) $\text{LOSS}\left(\theta,\ \hat{\theta}\right)=$

$$\frac{R_1^2\left(\hat{a}-a\right)^2}{4K_3P}I_{(S^*>S_0)}+\left\{-PK_3\left[\left(\overline{S}-S_0\right)^2-\left(\overline{S}-\hat{S}_0\right)^2\right]-PK_2\left(\hat{S}_0-S_0\right)\right\}$$

$$=I_{\{S^*=S_0\}}-R_1a\left(\hat{S}_0-S_0\right)I_{\{S^*=S_0\}}I_{\{S_0<\hat{S}_0\}}$$

where I takes values of 1 or 0,

$I_{\{\text{expression}\}}$	$=1$	expression true
	$=0$	otherwise.

As $\hat{a},\ \hat{S}_0$ are random variables, the loss function is also random. For given values of $S_0,\ a,\ \sigma_1^2,\ \sigma_2^2$, and for a given scatter $\boldsymbol{S}_T$ of the observations $S_1, \ldots, S_T$, the conditional expectation of the loss function is

(19 $\overline{\text{LOSS}}\left(\sigma_1^2,\ \sigma_2^2,\ a,\ S_0,\ \boldsymbol{S}_T\right)=E\left[\text{LOSS}\left(\theta,\ \hat{\theta}\right)/\sigma_1^2,\ \sigma_2^2,\ a,\ S_0,\ \boldsymbol{S}_T\right]$

$$=\frac{R_1^2}{4K_3P}\int_{-\infty}^{\infty}\left(\hat{a}-a\right)^2dN\left(a,\ V(\hat{a})\right)\ I_{\{S^*>S_0\}}$$

$$+\left\{-2K_3\overline{S}\int_{-\infty}^{\infty}\left(\hat{S}_0-S_0\right)dN\left(S_0,\ V\left(\hat{S}_0\right)\right)\right\}$$

$$+\left\{-2K_3\bar{S}\int_{-\infty}^{\infty}\left(\hat{S}_0 - S_0\right)dN\left(S_0,\ V\left(\hat{S}_0\right)\right)\right.$$

$$+PK_3\int_{-\infty}^{\infty}\left(\hat{S}_0^2 - S_0^2\right)dN\left(S_0,\ V\left(\hat{S}_0\right)\right)$$

$$\left.-PK_2\int_{-\infty}^{\infty}\left(\hat{S}_0 - S_0\right)dN\left(S_0,\ V\left(\hat{S}_0\right)\right)\right\}I_{\{S^*=S_0\}}$$

$$-R_I a\int_{S_0}^{\infty}\left(\hat{S}_0 - S_0\right)dN\left(S_0,\ V\left(\hat{S}_0\right)\right)I_{\{S^*=S_0\}}$$

$$=\frac{R_1^2}{4K_3P}V(\hat{a})\ I_{\{S^*>S_0\}}+\left\{PK_3V\left(\hat{S}_0\right)-\frac{R_1a}{\sqrt{2\pi}}\sqrt{V\left(\hat{S}_0\right)}\right\}I_{(S^*=S_0)}.$$

Let $G\left(\sigma_1^2,\ \sigma_2^2,\ a,\ S_0\right)$ be the joint prior distribution of $\sigma_1^2,\ \sigma_2^2,\ a,\ S_0$. Assume that these four variables are independent, and their marginal distributions are known[5]:

(a) $P_r\left(\sigma_1^2 = d_1\right) = P_r\left(\sigma_2^2 = d_2\right) = 1$

(b) $a \sim U\left[\delta_1,\ \delta_2\right]\ \delta_1 \le \delta_2 \le 0$

(c) $S_0 \sim U\left[\Delta_1,\ \Delta_2\right]\ 0 \le \Delta_1 \le \Delta_2$

where $d_1,\ d_2,\ \delta_1,\ \delta_2,\ \Delta_1,\ \Delta_2$ are known. Under these assumptions, the expectation of (19) will be

$$(20)\quad H\left(T,\ \boldsymbol{S}_T\right) = E\left\{\overline{\text{LOSS}}\left(\sigma_1^2,\ \sigma_2^2,\ a,\ S_0,\ \boldsymbol{S}_T\right)\right\}$$

$$=\frac{R_1^2}{4K_3P}\int_{\{S^*>S_0\}}V(\hat{a})\ dG\left(\sigma_1^2,\ \sigma_2^2,\ a,\ S_0\right)$$

$$+PK_3\int_{\{S^*=S_0\}}V\left(\hat{S}_0\right)\ dG\left(\sigma_1^2,\ \sigma_2^2,\ a,\ S_0\right)$$

$$-\frac{R_1}{\sqrt{2\pi}}\int_{(S^*=S_0)}a\sqrt{V\left(\hat{S}_0\right)}\ dG\left(\sigma_1^2,\ \sigma_2^2,\ a,\ S_0\right).$$

Let us now calculate the profitability of acquiring additional observations $(S_i,\ Y_i)$. For a sample of a given T let $\boldsymbol{S}_T^*$ be the scatter that minimizes (20). According to Yahav (1980) under assumptions (a)-(c)

(a) $\sigma_1^2 = \sigma_2^2$

(b) $S_0 \sim U[0, 1]$

(c) there is a continuum of observations,

the optimal spread, according to the Beta $\left(\frac{1}{3}, \frac{1}{3}\right)$ density, will be

$$f_{\frac{1}{3}, \frac{1}{3}}(S) = S^{\frac{1}{3}}(1-S)^{\frac{1}{3}} \frac{\left(\frac{5}{3}\right)!}{\left(\frac{1}{3}\right)!\left(\frac{1}{3}\right)!} \qquad \text{if} \qquad 0 \le S \le 1$$

$= 0$ otherwise.

The solution was adapted to our case, where the number of the observations is discrete, and further

(a) $\sigma_1^2 = d_1 \neq \sigma_2^2 = d_2$

(b) $S_0 \sim U[\Delta_1, \Delta_2]$;

$\frac{1}{3} d_1/d_2$ observations were taken in the vicinity of Δ_1, $\frac{2}{3} - \frac{1}{3} d_1/d_2$ in the vicinity of Δ_2, and $\frac{1}{3}$ uniform spread between the above two groups.

At this stage we have no proof that this spread will indeed minimize (20), and it is only an intuitive approximation to Yahav's result.

Finding the optimal spread is a complicated statistical problem whose analytical solution will not be discussed here. However, the suggested spread (hereafter "spread I") was empirically compared with two other alternative spreads of additional observations.

Assume that D hectares of potatoes are grown in the region where the (S_i, Y_i) observations were taken, with the same technology, soil, and climate conditions. Let n be the number of additional observations to be taken and $C_0(n)$ the cost of their acquisition. With $H(T, \boldsymbol{S}_T)$ describing the situation *a priori*, the expected value of additional information—to the potato growers of that region—from n observations with spread $\boldsymbol{S}^*_{T+n}$, is

$$\text{(21)} \qquad \text{EVSI}\ (n) = D\left[H(T, \boldsymbol{S}_T) - H(T+n, \boldsymbol{S}^*_{T+n})\right] - \boldsymbol{C}_0\ (n).$$

The optimal number of observations n^* can be determined by

$$\text{(22)} \qquad \text{EVSI}\ (n^*) = \max_n \text{EVSI}\ (n).$$

Table 9.2. Expected value of sample information [EVSI(n)] for the three alternative observations spreads (US$)

Number of observations (n)	3	6	9	12	15	18	21	24	27	30	33
EVSI (n), Spread I	14364	17185	22290	23658	23992	23898	22993	21058	19902	19600	19216
EVSI (n), Spread II	4610	8843	11277	12782	13614	14133	14398	14482	14424	14272	14032
EVSI (n), Spread III	6562	9149	11253	13550	14631	15467	16330	16738	17022	16742	16298

4.1. Empirical Results

The empirical application of the loss-function analysis to the Northern Negev potato-growing region is presented below. The aggregate expected benefits to the potato growers of the region from improved response data are calculated and the optimal sample size is determined.

Based on the empirical estimates of the response-function parameters, the following approximate values were assigned to d_1, d_2, δ_1, δ_2, Δ_1 and Δ_2:

$$d_1 = 0.1 \cong \hat{\sigma}_1^2; \qquad d_2 = 16 \cong \hat{\sigma}_2^2$$
$$\delta_1 = -1.30 = \hat{a} - 0.21; \qquad \delta_2 = -0.88 = \hat{a} + 0.21$$
$$\Delta_1 = 4.5 \cong \hat{S}_0 - 1.5; \qquad \Delta_2 = 7.5 \cong \hat{S}_0 + 1.5.$$

Accordingly, the marginal *a priori* distributions are

$$P_r\left(\sigma_1^2 = 0.1\right) = 1; \qquad P_r\left(\sigma_2^2 = 16\right) = 1.$$
$$a \sim U\left(-1.30,\ -0.88\right)$$
$$S_0 \sim U\left(4.5,\ 7.5\right).$$

The suggested spread I of additional observations was compared with two alternative spreads, spread II uniformly scattered in the range $S_0 = 4.5$ to $S_0 = 7.5$, and spread III uniformly scattered from $S_0 = 6.054\left(\hat{S}_0\right)$ to $S_0 = 20\left(\overline{S}\right)$. For P = 0.1 [US$/m^3], R_1 = 161 [US$/ton], D = 2000 [ha] (the Northern Negev, where this experiment was conducted, is the main production area of potatoes in Israel), and $C_0(n) = 260n$ [US$] (these expenses constitute US$130/observation of direct costs + US$130/observation due to the opportunity cost of the research personnel); the obtained values of EVSI(n) are presented in Table 9.2.

A number of observations and conclusions concerning the value of information may be gleaned from Table 9.2.

a) Spread I is substantially superior over the two other spreads for all values of n. Based on spread I, which is an intuitive approximation of Yahav's (1980) findings, the estimated optimal sample size is $n^* = 15$ and the expected value of additional information is US$23,992.

b) Since some approximations were used and since the underlying statistical theory is mainly asymptotic and assumes the use of large samples, the results, which are based on a medium-size sample, must be regarded as approximate. Their main value is that they enable us to learn the order of magnitude of EVSI and to draw operative conclusions about the benefit of additional sampling.

c) The present level of knowledge of the potato growers in that region is relatively high and any improvement in production due to a better knowledge of the response-function is expected to be relatively small.[6] Since a short-run optimization model was assumed, the results might be regarded as downward biased: actually, improved estimates of the parameters of the response function may contribute to the benefits of the farmers of the region more than one growing season. It should be expected, however, that a multiperiod optimization model (with soil salinity as a dynamic state variable), may yield a higher values of EVSI.

5. CONCLUSION

The estimation of the response function of a given crop to soil salinity and the calculation of the expected value of additional information on the parameters of this function are important steps in the process regarding irrigation with saline water under conditions of uncertainty.

A method for the estimation of a response function following Maas and Hoffman's (1977) specification was developed and the stochastic properties of the estimated parameters were discussed.

An optimization model for the determination of the optimal quantity of water from a given source needed to leach the soil was formulated. A loss function was constructed, its possible states were defined, and its expectation was derived.

Finally, the EVSI on the response-function parameters was calculated. The optimal sample size was determined with regard to the preferable spread of the additional observations.

The empirical results are obviously conditional on the quality of the physical data and incorporated assumptions and are therefore specific to the region under consideration. Although the procedures introduced are quite complex, once computer software is written, they should be easily extendable to other crops and regions at a relatively low cost.

There are at least two directions for possible extension of the analysis: (a) computing EVSI, in a long-run analysis, referring to the water-soil-crop farm system over a sequence of several irrigation seasons, taking into consideration the long-run soil-leaching process; (b) computing EVSI for a single crop within a multiculture farm framework, with several crop alternatives, several water sources differing in quality, quantity, and prices, and several fields differing in area and initial salinity (Feinerman, 1980). The analysis presented in this paper can serve as a building block in such extended analyses. Its main advantage seems to be in providing a conceptual and methodological framework with which to investigate the problem as well as an efficient tool for empirical analysis.

REFERENCES

Bernstein, L. Effects of salinity and soil water regime on crop fields in: "Salinity in Irrigation and Water Resources" (D. Yaron, Ed.). New York: Dekker. 1980.

Bickel, P. I. and I. A. Yahav. Some contributions to the asymptotic theory of Bayes solution, Z. *Wahrscheinlich Keitstheories Verw*, 2:257-276. 1969.

Bierhuizen, I. F. Water quality and yield depression. Institute of Land and Water Management Research. Wageningen, Netherlands. 1969.

Bresler, E. A model for tracing salt distribution in the soil profile and estimating the efficient combination of water quality and quantity under varying field conditions. *Soil Science*, 104:227-233. 1967.

Davis D. R. and W. M. Dvoranchik. Evaluation of the worth of additional data. *Water Resources Bulletin*, 7:700-707. 1971.

DeForges, I. M. Research on the utilization of saline water for irrigation in Tunisia. *Natural Resources*, 6:2-6. 1970.

DeGroot, M. H. "Optimal Statistical Decisions." New York: McGraw-Hill. Carnegie Mellon University. 1970.

Feinerman, E. Economic analysis of irrigation with saline water under conditions of uncertainty. (PhD Thesis). The Hebrew University of Jerusalem, Rehovot, Israel. 1980. (Hebrew)

Harville, D. A. Maximum likelihood approaches to variance component estimation and related problems. *Journal of American Statistics Association*, 72:320-340. 1977.

Klemes, V. Value of information in reservoir optimization. *Water Resources Research*, 13:837-850. 1977.

Maas E. V., and G. I. Hoffman. Crop salt tolerance-current assessment. *American Society of Civil Engineers Journal of Irrigation and Drainage Division, IR2*, 103:115-134. 1977.

Maddala, G. S. "Econometrica." New York: McGraw-Hill. 1977.

Maddock, T. Management model as a tool for studying the worth of data. *Water Resources Research,*. 9:270-280. 1973.

Nouri, A. K. H., I. V. Drew, Knudson and A. Olson. Influence of soil salinity on production of dry matter and uptake and distribution of nutrients in barley and corn: I. Barley. *Agronomy*, 62:43-48. 1970.

Polovin, A. Economic analysis of irrigation with saline water on a kibbutz farm (MSc Thesis) The Hebrew University of Jerusalem, Rehovot, Israel. 1974. (Hebrew)

Pratt, I. W., H. Raiffa, and R. Schlaifer. "Introduction to Statistical Decision Theory." New York: McGraw-Hill. 1965.

Quandt, R. E. Tests of the hypothesis that a linear regression system obeys two separate regimes. *Journal of American Statistics Association*, 55:324-340. 1960.

Ryan, I. G. and R. K. Perrin. Fertilizer response information and income gains: the case of potatoes in Peru. *American Journal of Agricultural Economics,* 56:337-343. 1974.

Sadan D. and M. Berglas. Personal communication. Ministry of Agriculture, Agricultural Extension Service. Beer Sheva, Israel. 1980.

Shalhevet, I. and L. Bernstein. Effects of vertically heterogeneous soil salinity on plant growth and water uptake. *Soil Science,* 106:85-93. 1968.

Yahav, I. A. Personal communication. The Hebrew University of Jerusalem, Rehovot, Israel. 1980.

D. Yaron, H. Bielorai, I. Shalhevet and Y. Gavish. Estimation procedures for response functions of crops to soil water content and salinity. *Water Resources Research,* 8:291-300. 1971.

Yaron, D. and A. Olian. Application of dynamic programming in Markov chains to the evaluation of water quality in irrigation. *American Journal of Agricultural Economics,* 55:467-471. 1973.

---

[1] Deleting the first two and the last three observation intervals eliminates the possibility $\hat{\sigma}_1 = 0$ or $\hat{\sigma}_2 = 0$ which would make (7) equal infinity for any $\hat{S}_0$ in these intervals.

[2] All the monetary values are at the January 1978 price level

[3] The observations, Y_i, are normally distributed, the prior density function is everywhere positive, and the loss function is proportional to the squared error. It is therefore asymptotically true that Bayes (1969) estimates (the parameter estimates that minimize the expected loss) are identical to the MLE (e.g., Bickel and Yahav, 1969).

[4] Based on (16) and (17), one may disregard the case $\hat{S}^* \leq S_0,\ S^* > S_0$. Hence, the loss function for (b) should be the same as the loss function for (a).

[5] Based on conversations with soil researchers and our *a priori* knowledge, we believe that it is possible to assign closed intervals to the unknown true parameters. Since we are not able to assign different probabilities to subset lengths of the interval mentioned above we assume prior uniform distributions.

[6] The optimal EVSI (US$23,992/region) found in Table 9.2 constitutes less than 1 percent of the total revenue of potato growers.

A MODEL FOR THE ANALYSIS OF SEASONAL ASPECTS OF WATER QUALITY CONTROL*

Dan Yaron
The Hebrew University of Jerusalem, Rehovot, Israel and
The University of Chicago, Chicago, Illinois, USA

1. INTRODUCTION

In the last decade, researchers have devoted considerable effort to the development of models capable of generating socially optimal solutions to problems of water quality management in river basins. Although the ideal formulation of such problems is to maximize net benefits from pollution abatement, because these benefits are hard to quantify, a second-best approach has been to minimize the social cost of achieving predetermined quality standards.

To this end, one group of researchers (e.g., Graves et al., 1969; Hass, 1970; Haimes, 1971; Hwang et al.; (1973), Herzog, 1976) has applied linear and nonlinear programming techniques. Such an approach has considerable value, in that it promotes understanding of the system and the interaction of its parts and can yield estimates both of alternative costs of various levels of water quality and of the trade-offs between variables of the system. A shortcoming of most of the mathematical programming approaches has been their deterministic framework; rather than reflecting the stochastic nature of the system, river flows and related parameters have been assumed constant.[1]

Models with semi analytical approaches like dynamic programming (e.g., Dysart and Hines, 1970; Hwang et al.; 1973) or with heuristic approaches like simulation (Davis, 1968) can easily accommodate stochastic elements, but their amorphous structures tend to obscure the economics of the system. Accordingly they lose some of the valuable economic insights gained by mathe-

*Permission to publish this chapter was granted by the Academic Press, Inc. The chapter was originally published under a similar title in the *Journal of Environmental Economics and management*, 6:140-155, 1979.

matical programming, which explicitly incorporates such economic relationships as shadow prices and coefficients of substitution.

Russell and Spofford (1972) and Russell (1973) did endeavor to blend the two approaches by using mathematical programming to analyze the socioeconomic part of the system (e.g. firms, wastewater treatment plants), along with a simulation model, to analyze the water quality and the ecological parts of the system. Although conceptually quite appealing, their approach seems to require considerable research resources for finding an optimal solution, in the iterative process employed, and in practice, its application is subject to the availability of such resources.

The strategy of our study is to separate the analysis into two stages. At the first stage a mathematical programming model, involving some simplifying assumptions, is applied to the whole system (including both its economic and ecological parts) in order to gain, with its aid, understanding of the interrelationships between its elements and arrive at approximate solutions and policy alternatives. At the second stage simulation is applied to test alternatives for decisions within a more accurate framework. Note that simulation alone, while efficient in testing alternative policies, has limited creativity and capability for the formulation of sound alternatives.

This chapter discusses the mathematical programming approach. We present an attempt to analyze water quality control problems with a multiseasonal mathematical programming model (following Dantzig's work on linear programming under conditions of uncertainty, 1955) into which elements of chance-constrained programming (Charles and Cooper, 1960) are incorporated. The idea is to capture phenomena of interseasonal variation by introducing into the model several "seasons" which are characterized by natural factors affecting water quality, and are represented in the model by "seasonal" parameters. The intraseasonal variation is incorporated into the analysis by formulation of "chance constraints." The specification of the variables as deterministic or stochastic is discussed later in the chapter. Note that "seasons" can be defined in terms of months rather than longer periods, thereby reducing the intraseasonal variation.

The idea of the multiseasonal approach arose following the scrutiny of empirical data related to water quality in Du Page County, northeastern Illinois. Seasonal variation in temperature is a well-known phenomenon, and there is no need for corroboration of this statement. The observed, apparently systematic variations in the intensity of river flows are shown in Table 10.1. Several examples of the pattern of seasonal variation in water quality are presented in Table 10.2.

Table 10.1. Relative monthly flows with respect to an "average month"[a] for several Illinois rivers

River	Gauging location	1	2	3	4	5	6	7	8	9	10	11	12
Du Page, West Branch	West Chicago	0.8	1.0	1.7	1.9	1.6	1.3	0.6	0.6	0.6	0.5	0.6	0.9
Fox	Algonquin	0.8	1.0	1.7	1.9	1.4	1.1	0.8	0.5	0.6	0.7	0.9	0.9
Des Plaines	Riverside	0.8	1.1	1.9	2.0	1.4	1.3	0.8	0.6	0.5	0.5	0.5	0.7
Vermillion	Pontiac	1.1	1.2	2.2	1.9	1.8	1.3	0.5	0.1	0.2	0.3	0.4	0.8
Kankakee	Momence	1.1	1.3	1.7	1.7	1.5	1.0	0.7	0.4	0.5	0.6	0.7	0.9

Computed from data in U.S. Geological Survey (1976).
[a] Computed as $(1/T)\ \Sigma_t\ [F_{mt} / (\Sigma_m F_{mt})/12]$, where F_{mt} is the flow in month m and year t, and T is the number of years for which data were available (generally 13).

Table 10.2. Relative monthly DO concentrations with Respect to an "average month"[a] for the Du Page River, Illinois

River	Gauging location	1	2	3	4	5	6	7	8	9	10	11	12
Du Page, West Branch	West Chicago	1.2	1.3	1.2	1.1	0.9	0.8	0.7	0.9	1.0	0.8	1.1	1.3
Du Page, West Branch	Naperville	1.3	1.5	1.2	1.0	1.0	0.6	0.8	0.9	0.8	0.9	1.0	1.3
Du Page, East Branch	Glenn Ellyn	1.1	1.5	1.7	1.6	0.9	0.7	0.9	0.6	0.7	0.5	1.1	1.4
Du Page, East Branch	Glendale Heights	1.2	1.1	1.5	1.5	0.8	0.5	0.6	0.7	0.7	0.6	1.3	1.5
Du Page, East Branch	Lisle	1.4	1.5	1.4	1.3	1.1	0.5	0.7	0.6	0.7	0.7	1.1	1.3
Du Page	Channahon	1.1	1.1	1.1	1.1	1.1	0.9	1.0	0.8	1.0	1.1	1.1	1.0

Computed from IEPA (unpublished) data.
[a]Computed as $(1/T)\ \Sigma_t\ [D_{mt} / (\Sigma_m D_{mt})/12]$, where D_{mt} is the DO concentration (mg/liter) in month m and year t, and T is the number of years for which data were available. The readings were taken sporadically over the years 1964-1975, with the number of observations for computing the monthly averages over T years ranging from 2 to 15 (in summer months).

Moreover, the Du Page River basin has potential for flow augmentation by storing storm runoff and floodwater. The years 1961-1967 record at least one annual flooding of the river's West Branch (U.S. Geological Survey, 1976) and show a range of one to eight. The available preliminary estimates (U.S. Army Corps of Engineers, 1976) of flood damage caused by the more severe floods in the years 1948-1972 in the West Branch of the Du Page River suggest that the total damage caused by those floods was between 5 and 8 million dollars.

The current water quality policy notably concentrates on relationships derived for the critical conditions, setting its regulations accordingly. Indeed a summary of the Illinois rules and regulations for effluent standards (NIPC, 1976, page 21) states that these standards "are based primarily on dilution available during low flow conditions which are assumed to be most critical. Critical low flows are defined as those associated with an occurrence which may be expected to extend over a 7-day period at a frequency of one in 10 years. The dilution ratio results from comparing the expected volume of effluent during the design year to the cited low flow." An examination of effluent permits issued by the EPA to industrial plants and wastewater treatment plants in the Du Page River basin in Illinois shows that effluent standards throughout the year are in fact constant, with no attempt to benefit from high-flow conditions. However, institutions and individuals seem willing to explore the possibilities of seasonal adjustments in the system.

2. OUTLINE OF THE MODEL

The model concerns a river divided into M reaches. The typical elements linked to the mth reach are industrial plants and a municipal wastewater treatment plant. The latter collects and treats wastewater discharged by the households and industrial plants linked to it, and the former have the additional option of discharging their pollutants directly into the river.

The unit of time is 1 year, divided into S seasons. A season is characterized by variables such as temperature, intensity of river flow, and the river's pollutant assimilative capacity.[2] The probability that season s will occur, $P(\theta_s)$, is assumed to be known and equals the expected number of days in the season divided by 365. The model is formulated as a two-stage mathematical programming problem (Dantzig, 1955) of minimizing the expected social cost of maintaining water quality at or above given standards with a predetermined probability level.[3] Population and industrial output levels are also given.

The decision variables *endogenously incorporated* into the model are industrial production technology (including on-site pollution abatement) and the levels of pollution emitted directly into the river and indirectly through the municipal plants. Incorporated into the model as *exogenous parameters* are decision variables related to the location, scale, and service area of the municipal treatment plants; land use regulations; and other measures concerning location.

Some of the endogenous decision variables are seasonally adjustable; their seasonal flexibility is supposed to reflect the real-life situation. In the first version of the model, these variables are limited to the level of treatment at the municipal treatment plants, but in the second version industrial production technology can also vary to take advantage of seasonal changes.

3. MATHEMATICAL FORMULATION

In the interest of brevity, but without loss of generality, only one pollutant[4] and a typical reach of the river will be considered. A transfer row and a transfer variable will represent the open-end link to the next reach.

The goal is thus to minimize z,

$$z = \sum_i \sum_j f_{ij}^D\left(X_{ij}^D\right) + \sum_i \sum_r f_{ir}^N\left(X_{ir}^N\right) + \sum_s P(\theta_s) g^s\left(t^s\right), \tag{1}$$

subject to

$$\begin{aligned}
&\sum_j X_{ij}^D + \sum_r X_{ir}^N \ge b_i, \\
&-\sum_i \sum_j a_{ij} X_{ij}^D + d^{Ds} = 0, \\
&-\sum_i \sum_r a_{ir} X_{ir}^N + t^s + d^{Ns} = H^s, \\
&-k \sum_i \sum_r a_{ir} X_{ir}^N + t^s \le k H^s, \\
&p^s d^{Ds} + p^s d^{Ns} \le \overline{Q}^s, \\
&-d^{Ds} - d^{Ns} + e^s = 0,
\end{aligned} \tag{2}$$

and

$$X_{ij}^D,\ X_{ir}^N,\ d^{Ds},\ d^{Ns},\ t^s,\ e^s \ge 0 \tag{3}$$

$i = 1, 2, \ldots, I$; $j = 1, 2, \ldots, J$; $r = 1, 2, \ldots, R$; and $s = 1, 2, \ldots, S$

where

X_{ij}^{D} = level of production output *i*, using technology *j*, involving direct discharge of the pollutant into waterways (units/day);

X_{ir}^{N} = level of production output *i*, using technology *r*, involving discharge of the pollutant into the municipal sewage system (unit/day);

$f_{ij}^{D}\left(X_{ij}^{D}\right)$, $d_{ir}^{N}\left(X_{ir}^{N}\right)$ = cost of production (financial outlay) of input *i* expressed as functions of X_{ij}^{D} and X_{ir}^{N}, respectively (US$/day);

b_i = predetermined level of daily output *i* (units/day);

a_{ij}, a_{ir} = quantity of the pollutant discharge per unit of product *i*, using technology *j* or *r*, respectively (lb/unit);

t = level of removal of the pollutant by the municipal treatment plant (lb/day);

$g(t)$ = cost of pollutant's removal as a function of *t* (US$/day);

d^{D}, d^{N} = quantity of pollutants discharged, respectively, directly into the waterways and indirectly through the municipal sewage system (lb/day);

H = predetermined discharge level of the pollutant by households into the municipal sewage system (lb/day);

k = maximum fraction of the pollutant removed at the municipal treatment plant using the present technology (scalar; ≤ 1);

$\bar{Q}$ = maximum reduction in water quality not violating β percent of the time the required water quality standard in the reach (mg/liter) (see the following section for the derivation of $\bar{Q}$)

p = "transfer coefficient," the level of reduction in water quality due to the discharge of one unit of the pollutant into the reach [(mg/liter)/lb];

e = quantity of the pollutant flowing into the following reach (lb/day).

Note that due to the natural digestion of the pollutant in the reach only a portion of *e*, say αe, $0 < \alpha < 1$ is assumed to enter the head of the following reach. The discrepancy between *e* and αe is due to the artificial segmentation of the river into discrete reaches.

The objective function represents the cost of production of the predetermined industrial outputs, plus the expected cost of treatment of wastewater by the municipal treatment plant over the *S* seasons. The second and third restrictions in (2) are definitions of d^{Ds} and d^{Ns} which represent the total quantities of the pollutant discharge, respectively, directly into the river and indirectly via the sewage system. The fourth restriction represents the upper limit on the level of removal of the pollutant. The water quality restriction (row 5) is derived from the relationship

(4) $P\{Q^{\text{nat},\, s} - p^s d^{Ds} - p^s d^{Ns} \geq Q^{\text{req},\, s}\} \geq \beta,$

where P denotes probability, $Q^{\text{nat},\, s}$ is the "natural" water quality when $d^{Ds} = d^{Ns} = 0$, $Q^{\text{req},\, s}$ is the water quality required in season s, and β is a prescribed probability level. We introduce some simplifying assumptions, namely, we refer to p^s as deterministic and to $Q^{\text{nat},\, s}$ as normally distributed with expectation, $E(Q^{\text{nat},\, s})$ and standard deviation σ^s. Under these assumptions (4) becomes equivalent to (5):[5]

$$(E(Q^{\text{nat},\, s}) - Q^{\text{req},\, s} - p^s d^{Ds} - p^s d^{Ns}) \,/\, \sigma^s \geq z_\beta,$$

(5) or

$$p^s d^{Ds} + p^s d^{Ns} \leq E(Q^{\text{nat},\, s}) - Q^{\text{req},\, s} - z_\beta \sigma^s \equiv \overline{Q}^s$$

where z_β is the value of the standard normal variable corresponding to the β probability level.

The assumption regarding the p's (= the transfer coefficients) as deterministic seems to be proper under conditions in which a large share of the variance of $(Q^{\text{nat},\, s} - p^s d^{Ds} - p^s d^{Ns})$ is due to the "natural" water quality. This is apparently the case in the West Branch of the Du Page River, for which the model was designed, and where approximately two-thirds of the Biochemical Oxygen Demand (BOD) loadings into the river originate in non-point sources (NIPC, 1977). If the above simplifying assumption is not valid, the variance of the whole linear expression $(Q^{\text{nat},\, s} - p^s d^{Ds} - p^s d^{Ns})$ should be considered and restriction (5) becomes nonlinear, implying that the solution of the problem becomes a formidable task.[6]

The last restriction in (2) and the variable e^s constitute the link to the following reach. Denoting the consecutive reaches by 1 and 2 the connection between the two reaches is expressed by

(6)
$$-{}^1 d^{Ds} - {}^1 d^{Ns} + {}^1 e^s = 0$$
$$-{}^1\alpha^s \cdot {}^1 e^s - \sum_u \sum_v {}^2 a_{u,\, v}\, {}^2 X^D_{uv} + {}^2 d^{Ds} = 0$$

In the following, the only aspect of relationship (6) that will be specifically mentioned is the linking shadow price γ^s of the pollutant (relating to the first row of (6)).

The cost functions f^D_{ij}, f^N_{ir} and $g^s(t^s)$ are assumed to be convex, a reasonable assumption for industrial and wastewater treatment plants with predetermined capacity and a given minimum removal level.[7] Reformulating $g^s(t^s)$ into $g^s(c, g^s)$in order to accommodate capacity (c) as a second argument may lead to loss of the convexity property. To avoid this difficulty, the model considers the capacity of the treatment plants as exogenously predetermined parameters. Accordingly, an empirical application of the model should test several runs with alternative configurations of treatment plants (location and capacity), but there is some solace in that few configurations will be *a priori* practical in general.

Let the dual variables corresponding to restriction set (2) be denoted by μ_i ($i = 1, 2, \ldots, I$), λ^{Ds}, λ^{Ns}, τ^{s}, δ^{s}, and γ^{s}, respectively. At the optimal solution of (1), (2), and (3) the following Kuhn-Tucker conditions are satisfied (since f_{ij}^{D}, f_{ij}^{N} and $g(t)$ are convex, and the restrictions are linear, the Kuhn-Tucker conditions are necessary and sufficient):

$$\mu_i^o - a_{ij}\sum_s \lambda^{oDs} \le \frac{\partial f_{ij}^{D}}{\partial X_{ij}^{oD}} \qquad \text{all } i \text{ and } j,$$

(7) $$\mu_i^o - a_{ir}\sum_s \lambda^{oNs} - ka_{ir}\sum_s \tau^{os} \le \frac{\partial f_{ir}^{N}}{\partial X_{ir}^{oN}} \qquad \text{all } i \text{ and } r,$$

$$\lambda^{oNs} + \tau^{os} \le P(\theta_s)\frac{\partial g^{s}}{\partial t^{os}}, \qquad \text{all } s,$$

$$\lambda^{oDs} + p^{s}\delta^{os} - \gamma^{os} \le 0,$$

$$\lambda^{oNs} + p^{s}\delta^{os} - \gamma^{os} \le 0 \qquad \text{all } s,$$

with

$$\mu_i^o,\ \lambda^{oDs},\ \lambda^{oNs},\ \gamma^{os} \ge 0,$$

and

$$\tau^{os},\ \delta^{os} \le 0$$

The superscript *o* denotes the variables' values at the optimal solution.

Rearranging the terms of (7) and referring to the absolute values of the non-positive Lagrangian multipliers we can proceed with the economic interpretation of the social cost of pollution. For $t^{os} > 0$:

(8) $$\lambda^{oNs} = P(\theta_s)\frac{\partial g^{s}}{\partial t^{os}} + \left|\tau^{os}\right|$$
$$= P\left((\theta_s)\frac{\partial g_s}{\partial t^{os}} + \left|\bar{\tau}^{os}\right|\right),$$

where $\left|\overline{\tau}^{os}\right| \equiv \left|\tau^{os}\right| / P(\theta_s)$ is the marginal cost of the technology constraint (at *k)* in season *s*, while $\left|\tau^{os}\right|$ is weighted by the probability $P(\theta_s)$. The weighted marginal social cost of pollution discharged in season *s* via the treatment plants is accordingly equal to the sum of the marginal cost of treatment plus the marginal cost of the technology constraint, with the sum weighted by $P(\theta_s)$. If we define $\overline{\lambda}^{oNs} \equiv \lambda^{oNs} / P(\theta_s)$, (8) becomes

$$(9) \qquad \overline{\lambda}^{oNs} = \frac{\partial g^s}{\partial t^{os}} + \left|\overline{\tau}^{os}\right|,$$

so that the outweighed marginal social cost of pollution, via the treatment plant in season *s*, is equal to the marginal financial cost of treatment plus the marginal cost of the technology constraint. (Note that $\overline{\tau}$ can be zero if the restriction is not binding.)

On the other hand,

$$(10) \qquad \lambda^{oNs} = p^s\left|\delta^{os}\right| + \gamma^{os}$$

or, dividing by $P(\theta_s)$,

$$(11) \qquad \overline{\lambda}^{oNs} = p^s\left|\overline{\delta}^{os}\right| + \overline{\gamma}^{os}.$$

The marginal social cost of pollution in a given reach in season *s* is the sum of the marginal cost of maintaining water quality in that reach plus the marginal cost of polluting the following reach.[8] Note that either δ^{os} or γ^{os} alone can be zero. If both are zero, $\lambda^{oNs} = 0$ is in contradiction to the assumption that $t^{os} > 0$

The marginal social cost of pollution due to direct waste dumping is

$$(12) \qquad \lambda^{oDs} = p^s\left|\delta^{os}\right| + \gamma^{os}$$

or

$$(13) \qquad \overline{\lambda}^{oDs} = p^s\left|\overline{\delta}^{os}\right| + \overline{\gamma}^{os}.$$

Considering the first row of (7) and rearranging the terms, we find that any process included in the optimal solution satisfies the following:

$$(14) \qquad \mu_i^o = \frac{\partial f_{ij}^D}{\partial X_{ij}^{oD}} + a_{ij} \sum_s \lambda^{oDs}.$$

Upon substitution of $\lambda^{oDs} \equiv P(\theta_s)\overline{\lambda}^{oDs}$, we find:

$$(15) \qquad \mu_i^o = \frac{\partial f_{ij}^D}{\partial X_{ij}^{oD}} + a_{ij} \sum_s P(\theta_s)\overline{\lambda}^{oDs}.$$

The marginal social cost of producing I is equal to the marginal financial outlay plus the expected value of the social cost of pollution due to direct discharges.

Similarly, for any industrial process producing product I and discharging wastewater into the municipal system:

$$(16) \qquad \mu_i^o = \frac{\partial f_{ir}^N}{\partial X_{ir}^{oN}} + a_{ir} \sum_s P(\theta_s)\overline{\lambda}^{oNs} - ka_{ir} \sum_s P(\theta_s)\left|\overline{\tau}^{os}\right|.$$

The interpretation of the last term in (16) becomes clear by substituting (9) into (16) and obtaining

$$(17) \qquad \mu_i^o = \frac{\partial f_{ir}^N}{\partial X_{ir}^{oN}} + a_{ir} \sum_s P(\theta_s)\left(\frac{\partial g^s}{\partial t^{os}} + (1-k)\left|\overline{\tau}^{os}\right|\right).$$

Recall that k is the maximum fraction of the pollutant removed at the waste treatment plant and that ka_{ir} is the increased treatable quantity of the pollutant in pounds due to one unit of X_{ir}^N in season s adjusted by a correction term $ka_{ir}\left|\overline{\tau}^{os}\right|$.

It should be noted that this section includes alternative formulations of the model, which may reflect different situations with respect to the flexibility of seasonal adjustments of other elements of the system. If, for example, all industries have the option of varying their technologies of production and on-site pollution abatement over the seasons, the objective function (1) would be reformulated to (18).

$$(18) \qquad z' = \sum_s P(\theta_s)\left[\sum_i \sum_j f_{ij}^D\left(X_{ij}^{Ds}\right) + \sum_i \sum_r f_{ir}^N\left(X_{ir}^{Ns}\right) + g^s\left(t^s\right)\right],$$

with the seasonal index s added to the X_{ij}^{D} and X_{ir}^{N} variables. In such a case the marginal social cost of production of product I may vary from one season to another. A further modification of the model may involve seasonal variation of the industrial output of some of the heavily polluting products.

4. IMPLICATIONS AND POSSIBLE APPLICATIONS OF THE MODEL

The importance of seasonal variability and the importance of policies intended to benefit from seasonal adjustments are issues to be examined within an empirical domain. A case study analysis will be presented in a later section. This section discusses some general ideas regarding the implications and the capability of the model.

4.1. Implications with Respect to User Charges and Pollution Taxation

Baumol and Oates (1971) have shown that by imposing proper user charges and pollution emission taxes it is possible to induce competitive industries to behave in a socially optimal manner. Estimates of such taxes have been derived in several empirical studies of water quality control (Hass, 1970; Haimes, 1971; Herzog, 1976; and others). Our model suggests that these taxes should be seasonally adjusted in order to reflect seasonal variations in the marginal social cost of pollution.

Assume that the marginal social cost of pollution in season s is $\overline{\lambda}^{oDs}$ and $\overline{\lambda}^{oNS}$ for "direct" and "indirect" emissions, respectively, and that the shadow price of the technological restriction on the level of the pollutant removal is $\left|\overline{\tau}^{oS}\right|$. Assume, further, that the producer of product i is charged for his emission $\overline{\lambda}^{oDs}$ and $(\overline{\lambda}^{oNS} - k\,|\,\overline{\tau}^{os}\,|)$ per unit of direct and indirect discharge. The optimizing (cost minimizing) problem of the ith industry can be expressed as follows:

Minimize

$$z_i = \sum_j f_{ij}^{D}\left(X_{ij}^{D}\right) + \sum_r f_{ir}^{N}\left(X_{ir}^{N}\right)$$

$$+\sum_s P(\theta_s)\left[\overline{\lambda}^{oDs}\sum_j a_{ij}X_{ij}^{D} + \left(\overline{\lambda}^{oNs} - k\left|\overline{\tau}^{os}\right|\right)\sum_r a_{ir}X_{ir}^{oN}\right] \quad (19)$$

subject to

$$(20) \qquad +\sum_j X_{ij}^D + \sum_r X_{ir}^N \geq b_i,$$

$$X_{ij}^D,\ X_{ir}^N \geq 0$$

For a solution of (19) and (20) to be optimal it is sufficient that the following Kuhn-Tucker conditions be satisfied:

$$(21) \qquad \begin{aligned} \phi_i &\leq \frac{\partial f_{ij}^D}{\partial X_{ij}^{oD}} + a_{ij} \sum_s P(\theta_s) \overline{\lambda}^{oDS} \\ \phi_i &\leq \frac{\partial f_{ir}^N}{\partial X_{ir}^{oN}} + a_{ij} \sum_s P(\theta_s) \left(\overline{\lambda}^{oNs} - k \left| \overline{\tau}^{os} \right| \right), \end{aligned}$$

where ϕ_i is a nonnegative number. But (21) is satisfied by letting $\phi_i = \mu_i^o$ as in (7). Accordingly, the optimal private solution of the industry producing the *i*th output is equivalent to the social optimum.

While the above argument is straightforward from the technical-formal point of view conceptually it raises several questions (as does the Baumol-Oates approach). It assumes that the tax-imposing authority and the private producers agree upon the relevant functions and parameters, a condition violated in the real world where numerous elements of the model are uncertain and variable. Under conditions of uncertainty Fishelson (1976) and Roberts and Spence (1976) have shown that a mixed water quality control system based on a combination of taxes and emission standards is more efficient than the pollution tax by itself. Although a discussion of the effectiveness of user charges and pollution taxes vs. emission standards is beyond the scope of this chapter, knowing the approximate values of the seasonal marginal costs of pollution and understanding the factors which determine them may be useful for policy decisions regarding water quality control, regardless of whether it were based on user charges and emission taxes, emission standards, or any combination thereof.

4.2. Other Possible Applications of the Model

With proper modifications, this model may also be used to analyze other issues in which seasonal variability is an important factor.

One application, for example, involves nonpoint sources of pollution and its effect on $Q^{\text{nat},\, s}$. Another application concerns an evaluation of possible savings in water pollution abatement costs, using river flow control by means of reservoirs (low-flow augmentation). Alternative policies of flow modification and/or instream aeration and their effects on the cost of pollution abatement can also be easily tested with this model. A still more ambitious and comprehensive study can involve the joint analysis of flood control, flow regulation, and water quality control, overcoming the problems of "dimensionality" by a decomposition approach.

5. CASE STUDY

The case analysis described here is confined to two reaches only, with one town, one treatment plant, and two industrial plants in each reach. The economic data are based on information from Northeastern Illinois Planning Commission (NIPC, 1977), Illinois Environmental Protection Agency, United States Environmental Protection Agency, Hydrocomp, Inc., and information obtained from wastewater treatment plants directly. The evaluation shows that they are representative of the conditions prevailing in the area. At the same time the two-reach system was synthetically constructed in order to contain enough elements of interest. Water quality parameters were taken from Hydrocomp, Inc. (1977) and from Dorfman and Jacoby (1972) and were somewhat arbitrarily adjusted to the size and flow intensity of the Du Page River, Western Branch. On the whole it seems that the following case example generally reflects the typical elements of the problem and the results of the analysis provide evaluation of the order of magnitudes involved.

In the application of the model, BOD is a pollutant of interest and the Dissolved Oxygen (DO) concentration is the measure of water quality. The treatment cost functions $g^s(t^s)$ were approximated by linear step functions corresponding to three levels of BOD removal: *L*1 ($\leq$90.5%), *L*2 (90.5-95.2%), and the third level, *L*3 (95.2-97.6%).[9]

Two seasons were distinguished, namely, "high flow" and "low flow," with respective probabilities of 0.75 and 0.25. The levels of $\bar{Q}^s$ (maximum reduction in water quality not violating the required water quality in the reach, 4 mg/liter DO, 99 percent of the time) were assumed to be 3.96 and 1.82 mg/liter DO, respectively for the high-flow and low-flow seasons. They were derived using relationship (5), with $E(Q^{\text{nat},\, s})$ and σ^s respectively 11.50 and 1.52 for the high-flow and 10.82 and 0.86 for the low-flow seasons, and $z_{0.99}$ = 2.326. The objective function was to minimize the total *variable cost* of pollution abatement, with the construction and the equipment of the treatment facilities predetermined. The variable cost includes operating and that part of maintenance cost which is avoidable on a seasonal adjustment basis. Note that all three-treatment levels were included in the model, and the question was what levels to operate during the two seasons. The essentials of the results are summarized in the following: a) During the high-flow season the first level (*L*1) of wastewater treatment only should be applied in the upper reach and both *L*1 and second level (*L*2) in the lower reach; b) During the low-flow season all three-treatment levels should be applied in both reaches; c) Accordingly, the marginal social cost of BOD removal varies significantly from one season to the next as shown below (units in Table 10.3 are dollars per pound);[10]

Table 10.3. Marginal social cost of BOD removal

	High-Flow	Low-Flow
Upper reach	0.59	1.27
Lower reach	0.66	1.59

d) The total variable cost per day of the two treatment plants is US$595 in the high-flow season and US$1,097 in the low-flow season; a difference of US$502/day; e) Abandoning the constant effluent standard policy and allowing for higher effluent standards in the high-flow season results in a potential cost reduction of US$137,046 per year (= US$502 × 273 days).

One way to evaluate the relative importance of the above saving potential is to relate it to a situation in which effluent standards and treatment levels are rigidly determined for the entire year, with reference to the critical low-flow conditions and no seasonal adjustments in the treatment levels. In such a case the total variable cost of treatment is US$400,405 per year (=US$1,097/day × 365 days), and the potential saving (US$137,046) due to the seasonal adjustment comprises 34 percent of the total variable cost.

6. CONCLUSION

The above results suggest that attention should be paid to the saving potential derived from seasonal adjustments in the treatment levels of wastewater treatment plants, in response to the flow conditions and the river's assimilative capacity. When considerable differences between low-flow and high-flow conditions prevail the design of the treatment plants should be adapted to seasonal flexibility. Such a design will emphasize the variable (operating) cost component and reduce the capital cost component (construction and equipment).

The model presented in this chapter can be useful in studying the trade-offs between these two components. To attain this goal alternative location and capacity configurations of treatment plants should be incorporated into parametric runs of the model. Alternatively, the model could be modified into a mixed-integer programming model to accommodate integer variables, representing location and capacity of treatment plants.

Finally, it should be noted that the model approximates some nonlinear relationships by linear functions. Accordingly, the analysis, with the aid of the model, can lead to a good understanding of the system and provide approximate solutions. These solutions should be tested later on by simulation within a more accurate framework. As previously mentioned, simulation alone has limited creativity and capability for the formulation of sound alternatives.

ACKNOWLEDGMENTS

This chapter was part of a project on the Economics of Water Quality in an Urban Setting funded by an NSF grant to the University of Chicago, with G. S. Tolley as the project leader. The efficient assistance of Bradley G. Lewis, graduate student in economics at the University of Chicago, is acknowledged. Helpful comments by D. Carlton, the participants of the University of Chicago workshop on Economics of Natural Resources, and an anonymous referee are appreciated.

REFERENCES

Baumol, W. J. and W. E. Oates. The use of standards and prices for protection of the environment. *Swedish Journal of Economics,* 73:42-54. 1971.

Charles, A. and W. Cooper. Chance constrained programming. *Journal of Institutional Management Science,* 6. 1960.

Dantzig, G. B. Linear programming under uncertainty. *Management Science,* 1:197-206. 1955.

Davis, R. K. "The Range of Choice in Water Management: A Study of Dissolved Oxygen in the Potomac Estuary." Johns Hopkins University Press. Baltimore, MD. 1968.

Deininger, R. A. Water quality management: the planning of economically optimal pollution control systems (PhD Thesis). Northwestern University. Evanston, IL. 1965.

Dorfman, R. and H. D. Jacoby. An illustrative model of river basin pollution control in: "Models for Managing Regional Water Quality" (D. Dorfman, H. D. Jacoby, and H. A. Thomas, Jr., eds.). Harvard University Press. Cambridge, MA. 1972.

Dysart, III, B. D. and W. W. Hines. Control of water quality in a complex natural system, IEEE *Transactions Systems Science Cybernet,* SSC6:322-329. 1970.

Fishelson, G. Emission control policies under uncertainty. *Journal of Environmental Economics and Management,* 3:189-197. 1976.

Frankel, R. J. Economic evaluation of water quality: an engineering economic model for water quality management. *Research Report,* 65-B. University of California, Berkeley, CA. 1965.

Graves, G. W., G. B. Hatfield and A. Whinston. Water pollution control using by-pass piping. *Water Resources Research,* 5:13-47. 1969.

Haimes, Y. Y. Modelling and control of the pollution of water resources systems via multilevel approach. *Water Resources Bulletin,* 10: 93-101. 1971.

Hass, J. E. Optimal taxing for the abatement of water pollution. *Water Resources Research,* 6:353-365. 1970.

Herzog, Jr., H. W. Economic efficiency and equity in water quality control: effluent taxes and information requirements. *Journal of Environmental Economics and Management*, 2:170-184. 1976.

Hwang, C. L., J. L. Williams, R. Shojalashkari, and F. T. Fan. Regional water quality management by the generalized reduced gradient method. *Water Resources Bulletin*, 9:1159-1180. 1973.

Hydrocomp, Inc. Water quality simulation of the Kishwaki River. Personal communication. Hydrocomp, Inc. Chicago, IL. 1977.

Marks, D. H. Water quality management in "Analysis of Public Systems" (A. W. Drake, R.L. Keeney, and P. M. Morse, eds.). MIT Press. Cambridge, MA. 1972.

NIPC (Northern Illinois Planning Commission). The framework for 208 planning in northeastern Illinois. Staff Paper, No.6. Chicago, IL. 1976.

NIPC. Du Page River Basin: Existing water quality conditions. Progress Report. Chicago, IL. Mimeo. 1977.

Parker, H. W. "Wastewater Systems Engineering." Prentice-Hall. Englewood Cliffs, NJ. 1975.

Roberts. M. J. and M. Spence. Effluent charges and licenses under uncertainty. *Journal of Public Economics*, 5:193-208. 1976.

Russell C. S. and W. O. Spofford, Jr. A quantitative framework for residuals management decisions in: "Environmental Quality Analysis: Theory and Method in Social Sciences" (A. V. Kneese and B. T. Bower, eds.). Johns Hopkins University Press. Baltimore, MD. 1972.

Russell C. S. Application of microeconomic models to regional environmental quality management. *American Economic Review*, 63:236-243. 1973.

U. S. Army Corps of Engineers. Plan of study, Chicago-South End of Lake Michigan urban water damage study. U. .S. Army Corps of Engineers. Chicago, IL. Mimeo. 1976.

U. S. Geological Survey. Water data for Illinois. U. S. Geological Survey. Washington, DC 1962-1976.

---

[1] See Marks (1972) for a similar view.

[2] An operational characterization of a season is presented below in the mathematical formulation of the model.

[3] The first stage relates to decisions with respect to variables, which remain constant throughout the whole year; the second stage relates to decision variables, which may be seasonally varied. For details see the mathematical formulation in the following section.

[4] Introducing additional pollutants raises empirical questions concerning specification of their interrelationships and will be discussed elsewhere.

[5] Note that p^s and $Q^{nat,\,s}$ in equation (5), and α^s in equation (6) fully characterize the natural conditions in season, s.

[6] A different formulation of "chance constraints" for a similar problem can be found in Deininger (1965). His formulation leads to nonlinear relationships; no attempt to solve an empirical problem is presented.

[7] See Frankel (1965).

[8] Referring to (6) we have, for ${}^1e^{os} > 0$, ${}^1\gamma^{os} = {}^1\alpha^{s}\,{}^2\lambda^{oDs}$

[9] These levels correspond to 20, 10 and 5 mg/liter BOD in the effluent of the treatment plants. $L1$ roughly corresponds to what is generally called secondary treatment, and $L2$ and $L3$ corre-

spond to sub-segments of tertiary or advanced treatment (Parker, 1975). The lower bound for treatment process *L*1 was not specified because it was redundant.

[10] Note that the model and the results refer to variable cost only.

11

TREATMENT OPTIMIZATION OF MUNICIPAL WASTEWATER AND REUSE FOR REGIONAL IRRIGATION*

Ariel Dinar and Dan Yaron
Center for Agricultural Economics Research, and
The Hebrew University of Jerusalem, Rehovot, Israel

1. INTRODUCTION

Municipal wastewater warrants increased attention as a potential environmental pollution and a possible irrigation water source. Under certain conditions, use of municipal effluent (treated wastewater) for irrigation is an effective means for wastewater removal. Using wastewater for irrigation of certain crops (Table 11.1) allows a less stringent treatment level in comparison to disposing of the wastewater to lakes and rivers to be utilized later and may thus alleviate environmental problems. It also has the advantages of providing extra water for farmers who may use the wastewater.

Municipal wastewater is generally treated in Western Europe and part of the United States (Messer, 1982; Asano and Mandancy, 1982) for discharge to streams and lakes, which might ultimately be used as sources of drinking water. Accordingly, the professional literature concentrates primarily on this issue and addresses: (1) the choice of a treatment facility that fulfills given health and environmental requirements at minimal cost and (2) the related treatment cost-sharing scheme among the polluting agents such as domestic and industrial users (Dorfman, 1972; Giglio and Wrightington, 1972; Papke et al., 1977; Loehman et al., 1979; Nakamura and Brill, 1979; Rinaldi et al., 1979).

Irrigation with effluent is a rather recent practice, and therefore the literature in this field is not as extensive as the literature dealing with the disposal of municipal wastewater to lakes and rivers. In arid and semiarid parts of the

*Permission was granted to publish this chapter by the American Geographical Union. The chapter was originally published in *Water Resources Research*, 22(3):331-338, March, 1986.

world like southern California, Texas, and Israel, much effort is being devoted toward coordinating effluent quality with agricultural crop requirements and toward adapting environmental quality regulations for using effluent for irrigation as a cost effective outlet for wastewater (Feigin et at., 1977; Tahal, 1978; Moore et al., 1984; Victurine et al., 1984; Goodwin et al., 1984; California State Water Resources Control Board, 1984).

This paper deals with a regional approach for municipal wastewater management through treatment and irrigation, subject to strict public health regulation aimed at preserving environmental quality. A regional wastewater treatment system and its distribution to farms within a region offers economic advantages to the potential participants, but it requires the establishment of a special regional organization. A region involved in such endeavor faces the following inter-related problems: (1) determination of the appropriate regional boundaries with due consideration to treatment plant capacity as well as the capacity and layout of the wastewater and effluent conveyance systems; (2) determination of the wastewater treatment level, (3) allocation of the effluent to the farms within the region, (4) the selection of optimal cropping patterns, (5) cost allocation to the participants, and (6) level of government subsidy, if needed.

A mathematical programming model, which includes these components, is formulated. The model is applied to the Ramla region, which includes one town and several farms on the coastal plain of Israel. The cost allocation among the participants is considered by Dinar et al. (1986). As will be shown later, the regional optimization problem can be separated from the cost allocation problem.

2. A MODEL FOR REGIONAL OPTIMIZATION OF WASTEWATER TREATMENT AND REUSE

The objective is to maximize the region's income subject to a given supply of wastewater, health regulations, the capability of the farms to utilize the effluent, subject to their land and other resource endowments, and the prevailing price system and technology.

The following entities, or "players" (using game theory semantics), are involved: (1) the municipal authority(s) that delivers the effluent (effluent supply), (2) the farmers interested in using effluent in irrigation (effluent demand), and (3) a public organization, such as the government or a regional authority which serves public interests (e.g., environmental quality) and which can control the related activities via regulations and subsidized financing.

Apparently, economic potential exists for regional cooperation in the treatment of wastewater and the use of effluent in irrigation. The farmers might be able to increase their irrigated acreages and benefits. The treatment cost could rise, as the result of a treatment level higher than required by health regulations for discharge to the sea, but a share of the cost would be borne by the farmers and by a government subsidy if needed. Environmental considerations and freshwater savings may provide the motivation for the subsidy.

The economic analysis refers to a one-year period, with all long-run costs and revenues expressed on an annual basis. It does not account for the effect of present irrigation decisions on the future from the standpoint of salt accumulation in the soil because with reference to the particular region studied, it is not significant due to salt leaching by winter rainfall.

Table 11.1. Effluent quality requirement for major crops

	Crop Group		
	A	B	D
BOD level, (mg/L)	<60	<35	<15
Coliforms, (Bacteria/100 mL)	...	<250	<12[a]
Treatment Level (index *j*)	1	2	3
Crops	cotton sugar beet seed crop cereal hay crops silage crops	fodder crops peanuts olives dates almonds citrus pecans other fruits with inedible peels deciduous fruit crops irrigated under canopy fruits and vegetables for canning vegetables consumed after cooking vegetable eaten without peels	unrestricted irrigation

Source: *Water Commission* (1978)
[a]For 80 percent of the samples

The model deals with one urban authority and several farms and incorporates government environmental quality regulations. It neglects seasonal differences in wastewater quality and assumes only one possible treatment level throughout the year; this level is chosen in order to maximize the regional income.

3. ELEMENTS OF THE MODEL

3.1. The Town and the Treatment Plant

Municipal wastewater supply and quality are viewed as predetermined exogenous variables. For each month, the following balance equation holds:

$$(1) \quad \overline{S}_t = D_t + S_t \qquad t = 1, \ldots, 12$$

where

$\overline{S}_t$ = municipal wastewater supply in month t, m^3;
D_t = quantity of wastewater discharged after necessary treatment in month t (discharged to the sea through the wadi (a riverbed, dry most of the year)), m^3;
S_t = quantity of wastewater diverted to a pretreatment storage in month t, m^3.

The cost of wastewater discharge is *d* (in dollars per cubic meter), and it is lower than the cost per unit of treatment required for irrigation use. Discharge costs include minimum wastewater treatment (required by law) and the conveying expense to the discharge site. The supply of wastewater is continuous over the year, while the agricultural demand for effluent is mainly restricted to the summer months. To match agricultural demand, wastewater is stored in a pretreatment reservoir until needed. Although some quality improvement (oxidation) does take place in the stored wastewater during winter, its extent with regard to the conditions studied is not known, and for the purpose of this study it will be considered as negligible. In general, there is some conflict as to whether wastewater should be stored before treatment or after treatment, as effluent. In the first case, there is the benefit of quality improvement before treatment, which reduces treatment costs; on the other hand, this implies that the treatment will take place at the appropriate time or effluent use, and a larger treatment plant is required. In the second case, the opposite holds. In this study, pretreatment storage in a large abandoned quarry conveniently located in the region was assumed with a small operational reservoir for the treated wastewater.

Denoting by *t* the index of the month, we distinguish between the group of the fall-winter months *w*, $w = \{9, 10, 11, 12, 1, 2, 3, 4\}$, and the peak summer months *t*, $t = 5, 6, 7, 8$. The fall-winter months will be treated as one group, *w*. We will also define $T = \{5, 6, 7, 8\} \cup w$.

The fall-winter storage treatment balance is

(2) $$U_{w5} + \sum_{j=1}^{3} Z_{wj} = (1-\alpha)\sum_{t \in w} S_t$$

where

α = loss coefficient due to evaporation and infiltration during winter-fall;
j = index of treatment level for use in irrigation; three discrete treatment levels are assumed (j = 1, 2, 3); j = 0 denotes the discharge option before storage, without a treatment plant;
U_{w5} = quantity of wastewater transferred from fall-winter storage during May $(t = 5)$, m^3;
Z_{wj} = quality of wastewater designated for treatment at level j during fall-winter, m^3.

The balance equation of storage and treatment in peak month t is

(3) $$U_{t,t+1} + \sum_{j=1}^{3} Z_{tj} = (1-\beta)[S_t + U_{t-1,t}] \qquad t = 5, 6, 7, 8$$

where

$U_{t,t+1}$ = quantity of stored wastewater transferred at the end of month t to month $t + 1$, m^3;
Z_{tj} = quantity of wastewater removed from storage to the treatment plant for treatment level j in month t, m^3;
β = loss coefficient for peak months.

Defining q_j as the maximal quantity or wastewater treated in the plant during any of the peak months t at treatment level j:

$$q_j = \max Z_{tj} \qquad t = 5, 6, 7, 8 \qquad j = 1, 2, 3$$

In the mathematical programming model applied, this relationship is expressed as

(4) $$q_j \geq Z_{tj}$$

Note that q_j is linked to a negative coefficient in the objective function which maximizes the region's income; therefore it should be as small as justified.

The region can choose one or the three levels or treatment and/or the discharge option (j = 0, 1, 2, 3). This is expressed by

(5) $$\sum_{j=0}^{3} \delta_j = 1$$

with

(6) $$\delta_j = 0,\ 1 \qquad j = 0, 1, 2, 3$$

(7) $$q_j \le \delta_j M \qquad j = 1, 2, 3$$

with M being an arbitrary large number. Thus, in the case of j = 1, 2, 3 for $\delta_{j*} = 0$, $q_{j*} \le 0$ which implies $q_j = 0$, for $j \ne j*$. For $\delta_{j*} = 1$, $q_{j*} \le M$; with M being sufficiently large, q_{j*} is practically unlimited and continuous for each j (j = 1, 2, 3), which is chosen. Notice that for j = 0 the town discharges all its wastewater; no treatment plant is established. Equation (7) allows, for $j \ne 0$, the treatment of part of the effluent and discharge of the other part. The model assumes certainty in the agricultural demand for water; therefore the quantity of treated effluent is determined by the demand for water in the irrigation season; the remainder is discharged after minimal treatment, which costs d (in dollars per cubic meter). The quantity which was stored (after losses) is being treated in a treatment plant and devoted to irrigation.

For each month, there are balance equations of supply (Z_{tj}) and uses of treated effluent:

(8) $$Z_{tj} \ge \sum_{n=2}^{N} R_{tj}^{n} \qquad j = 1, 2, 3 \qquad t \in T$$

where N is the number of "players" in the region (N is the number of farms plus one, the town) and R_{tj}^{n} is the effluent amount (in cubic meters) at treatment level j acquired by farm n = (n = 2, 3, ..., N) during month t. (Index $n = 1$ stands for the town.)

The plant's capacity Q_j (in cubic meters per month) is determined by q_j, defined above (4), with the addition of a safety factor γ:

(9) $$Q_j = \gamma\, q_j \qquad j = 1, 2, 3$$

The wastewater treatment cost is a nonlinear function (Loehman et al., 1979; Dinar, 1984).

A rate r of government subsidy of the treatment cost and conveying capital cost is assumed so that the actual cost function to the region is

$$(10) \quad P_j = (1-r)F_j(Q_j) \qquad j = 1, 2, 3$$

The following estimate (Dinar, 1984) was used:

$$F_j(Q_j) = 2006 Q_j^{\,0.633} E_j^{\,-0.094}$$

where E_j is the index of treatment level represented by the percentage of biochemical oxygen demand (BOD), remaining in effluent out of the pretreatment original 400 mg/liter. For $j = 1, 2, 3$, these percentages are 15, 8.75, and 4 (60, 35, and 15 mg/liter), respectively. The nonlinear cost function (10) is incorporated into the programming model by a separable programming routine (Control Data Corporation, 1977).

The last equation of the town expresses the cost of transporting wastewater from the town to the treatment plant with the site of the treatment plant being predetermined. Wastewater transport cost to the plant comprises (1) capital cost and (2) variable cost (mainly energy). Specifically, the following conveying cost function was assumed:

$$(11) \quad m^1 = (1-r)B^1(K^1) + v^1 \sum_{t=1}^{12} S_t$$

where

- m^1 = overall annual cost of conveying wastewater from the town to the reservoir, in U.S. dollars ;
- B^1 = capital cost as a function of K^1, dollars;
- v^1 = cost of energy per unit of wastewater conveyed from the town to the storage, U.S. dollars/m^3;
- K^1 = town's maximal periodic supply, m^3.

K^1 is determined by $K^1 = \max S_t$ ($t = 5, 6, 7, 8$), which is formulated in the programming model as

$$(12) \quad K^1 \geq S_t \qquad t = 5, 6, 7, 8$$

K^1 has a negative coefficient in the objective function, which is being maximized.

When the town operates alone, its goal is to minimize the treatment and conveying costs:

$$f^1 = d\sum_{t=1}^{12} D_t + \sum_{j=1}^{3} P_j + m^1$$

Within the regional framework, which is aimed at regional optimization, the town increases its expenses, assuming that the farmers will contribute their share. The above cost function multiplied by (-1) is one component in the regional objective function, which is maximized.

3.2. The Farms

The farms in the region differ in their production factors, their technology, and their cropping pattern as they relate to the possible regional treatment plant. Their major characteristics are presented in this section along with the relevant components of the programming model.

We denote the farm's group by G. It consists of N - 1 farms, $G = \{2, 3, \ldots, N\}$. Farm n $(n \in G)$ is characterized by L^n land sections and Y^n crop alternatives. Each land section of each farm can be irrigated with effluent, but due to sanitary regulations, it is not possible to irrigate the same land section with both effluent and freshwater during a season, nor to shift, over the years, from effluent to freshwater irrigation, unless special sanitary prevention measures are taken and permission is granted. Each farm can freely transfer freshwater among its land sections as long as its water quota allotments are not exceeded.

The farms can also install irrigation equipment on their nonirrigated areas and grow their irrigated crops. Each farm may have two out of four types of irrigation water at its disposal: $k = 1, 2, 3, 4$, namely effluent at treatment level $k = 1, 2, 3$, and freshwater $k = 4$, according to the farm's quota allotment. Recall that only one level of treatment of effluent throughout the season is possible.

Farm n's productive capacity is described by the following equations and inequalities:

Land use

$$(13) \qquad I_l^n \geq \sum_{y \in Y^n} \sum_{k=1}^{4} X_{ylk}^n \qquad l = 1, \ldots, L^n$$

where X_{ylk}^n is the area (in hectares) of crop y grown in land section l and irrigated with water of quality k by farm n, $l = 1, \ldots, L^n$; $k = 1, 2, 3, 4$; $y = 1, \ldots, Y^n$.

For $k = 1, 2, 3$, water quality used in irrigation is equal to quality of wastewater treated up to the jth level ($j = 1, 2, 3$); $k = 4$ denotes freshwater supplied from the conventional water system. I_t^n is the area of farm n's section l (in hectares).
Water use balance

$$(14) \qquad \overline{W}_{tk}^n \geq \sum_{l=1}^{L^n} \sum_{y \in Y^n} A_{yltk}^n - R_{tj}^n$$

$$t \in T \qquad k = 1, 2, 3, 4 \qquad j = 1, 2, 3 \qquad n \in G$$

where A_{yltk} is the irrigation water amount (in cubic meters) of quality k applied in period t per hectare of activity y in farm n's section l and $\overline{W}_{tk}^n$ is the farm n's irrigation water supply and use (in cubic meters) of quality k in period t. Effluent supply and use (equation (8))

$$Z_{tj} \geq \sum_{n \in G} R_{tj}^n \qquad j = 1, 2, 3 \qquad t \in T$$

Farm n's site in the region determines effluent conveying costs from the treatment plant (whose site is given) to its fields. The capacity of the effluent conveying system is determined by the maximum periodic effluent supply that must be transported from the plant to these fields:

$$(15) \qquad \sum_{j=1}^{3} R_{tj}^n \leq K^n \qquad t \in T \qquad n \in G$$

where K^n is farm n's maximal periodic effluent supply (in cubic meters). Note that (15) is equivalent to $R_{tj}^n \leq K^n$ for $j = 1, 2, 3$.

Conveying cost function from the plant to farm n's fields is

$$(16) \qquad m^n = (1-r)B^n(K^n) + v^n \sum_{t \in T} \sum_{j=1}^{3} R_{tj}^n$$

where v^n is the energy costs (in U.S. dollars per cubic meter) of conveying effluent from the treatment plant to farm n.

The characteristics of m^n are identical to those of m^1, which were already discussed in the section concerning the town. The m^n function is also treated with the aid of separable programming. We assume that the energy component in conveying costs depends linearly on the amount of effluent.

Additional restrictions for farm n are represented as follows:

(17) $$\boldsymbol{H}^n \, \tilde{\boldsymbol{x}} \leq \boldsymbol{b}^n$$

where

H^n = matrix of input factors (other than water) for farm *n;*

x^n = vector of activities not using water and not generating income, to farm *n*;

$\boldsymbol{b}^n$ = vector of restrictions not related to irrigation, specific to farm *n.*

The objective of farm *n* is to maximize f^n:

(18) $$f^n = \sum_{l-1}^{L^n} \sum_{y \in Y^n} \sum_{k=1}^{4} C^n_{ylk} X^n_{ylk} - m^n$$

where C^n_{ylk} is the gross income (in dollars per hectare) for activity unit y in farm n's land section l irrigated by water of quality k (market value net of marketing cost minus variable cost not including freshwater cost). The regional objective function f^N is composed of such N - 1 individual functions and the town's effluent treatment cost.

The regional goal is to maximize f^N:

(19) $$f^N = -f^1 + \sum_{n \in G} \left[\sum_{l=1}^{L^n} \sum_{y \in Y^n} \sum_{k=1}^{4} G^n_{ylk} X^n_{ylk} - m^n \right]$$

subject to restrictions (l)-(17) described above and the non-negativity restriction on the decision variables. The model makes it possible to determine the amount of regional income when the town's wastewater is used for irrigation within the framework of regional cooperation.

The decision variables in this model are *j*, treatment level of effluent or the discharge option, K^n, capacity of the conveying system of wastewater or effluent to/from participant *n* (*n* = 1, 2, 3, 4), Q^j, treatment plant capacity for treatment level *j* (note that the capacity and treatment level are determined simultaneously), and, X^n_{ylk}, level of activity *y* in land section *l* of producer *n,* irrigated by water of quality *k.* Decision variables determined exogenously to the model are the site of the treatment plant and *r,* the rate of government subsidy for treatment and conveyance.

3.3. Cooperative and Non-cooperative Solutions

As stated, f^N denotes the regional gross income when the farms and the town in the region cooperate $(N = \{1\} \cup G)$, and f^n denotes the gross income or the cost generated by the *n*th participant when acting independently ($n = 1, 2, \ldots, N$). A necessary condition for regional cooperation is that

(20) $$f^{N\circ} > \sum_{n=2}^{N} f^{n\circ} - f^{1\circ}$$

where o denotes the optimal values of f^N and f^n.

Other conditions for cooperation deal with the related cost benefit allocation schemes among the farms and the town. These are discussed by Dinar et al. (1986).

The model is able to solve for both the cooperative and the noncooperative situation. In the first situation the town and all the farms in the region cooperate to treat the municipal wastewater and use the effluent for irrigation (using game theory semantics a "grand coalition" is formed); in the second situation the town and each of the farms operate independently. In this case, the town disposes the wastewater, and the farms use their freshwater quota allotments only.

Intermediate situations, namely, cooperation among the town and some of the farms, are also possible ("partial coalitions" in game theory terms). Note that any cooperation, a grand or a partial coalition, is possible only if the town participates and supplies effluent.

The model includes the options to treat or not to treat wastewater for irrigation, and for each farm, to use or not to use the effluent. This formulation leads to the following result:

(21) $$\xi^{1\circ} \le f^{1\circ} \qquad \xi^{1\circ},\ f^{1\circ} < 0$$

$$\xi^{n\circ} \ge f^{n\circ} \qquad \xi^{n\circ},\ f^{n\circ} > 0 \quad n = 2, 3, \ldots, N$$

where $f^{1\circ}$ and $\xi^{1\circ}$ are the town's treatment cost in the non-cooperative and the cooperative optimal solutions, respectively; $f^{n\circ}$ and $\xi^{n\circ}$ are the *n*th farm income in the noncooperative (= no use of effluent for irrigation) and cooperative (= reuse of effluent) solutions, respectively.

Relationship (21) holds because the objective function (19) is to maximize the region's income, which is the sum of the town's treatment cost and of the $(N-1)$ farms' incomes. If wastewater treatment is not profitable from the region's standpoint (its objective function), i.e., $\delta_1 = \delta_2 = \delta_3 = 0$, then the programming problem (1)-(17) and (19) becomes a set of *N* independent programming problems and the solution represents the individualistic solutions of the town and each of the farms, with $\xi^{n\circ} = f^{n\circ}$ for all *n*.

If the use of effluent for irrigation is profitable from the region's point of view for any subset of farms $s \subseteq G$, then a coalition s^* between the town and that subset will be created, $s^* = s \cup \{1\}$, with the other farms $(N - s)$ acting independently. (In terms of the model solution this is obtained by allocating effluent to the farms $n \in s^*$.) The optimal value of the regional objective function is

$$f^{N\circ} = f^{\{s^*\}\circ} + \sum_{n \notin s^*} f^{n\circ}$$

where $f^{\{s^*\}\circ}$ is the s^*th coalition income. Notice that

$$f^{\{s^*\}\circ} = \sum_{n \in s^*} \xi^{n\circ} \geq \sum_{n \in s^*} f^{n\circ} \qquad \xi^{1\circ},\ f^{1\circ} < 0$$

and $\xi^{n\circ} \geq f^{n\circ}$ for all $n \in s \in s^*$. The value of the regional objective function is not optimized if for some farm n, $n \in s^*$, $\xi^{n\circ} < f^{n\circ}$. In game theory terms, "individual rationality" holds for each farm. The difference $\sum_{n \in s}\left(\xi^{n\circ} - f^{n\circ}\right)$ can be used for the compensation to the town for its additional treatment cost. Notice that $\sum_{n \in s}\left(\xi^{n\circ} - f^{n\circ}\right) \geq f^{1\circ} - \xi^{1\circ}$ and the compensation is possible.

4. EMPIRICAL ANALYSIS

4.1. Data and Description of the Region

The empirical analysis is being applied to a real case in the Ramla region on the coastal plain of Israel. The regional system consists of three farms, a town, and a wastewater treatment plant.

The town supplies wastewater of a given quality (400 mg/l BOD) and a constant quantity (100,000 m^3/month). The cost of disposing wastewater to non-farm sites (in this case, the sea) is 0.30 U.S. dollars/m^3. (All monetary values are constant October 1980 dollars.) If a regional treatment plant is set up, the town operates the plant with the understanding that the effluent will be distributed among the agricultural producers, who will purchase the effluent and pay at least, for the additional treatment costs.

Conveyance and treatment cost functions were described in the model section (for detailed data see Dinar, 1984). Wastewater storage loss coefficient for fall-winter (α) is 16 percent and for each peak month (β) is 4 percent (Berezik, 1982). The sanitary requirements for effluent quality are presented in Table 11.1.

Table 11.2. Basic data for representative crops

Crop	Loss Coefficient		Variable Cost Not Including Water and Labor, U.S.$ /ha	Labor, Days/ha		Yield Price Minus Marketing Cost U.S.$/t	Water, m^3/ha					Yield with Fresh Water, t/ha
	b	*a*		Culti-vating	Harvest		May	June	July	Aug	Annual	
Cotton, 3 irrigations by sprinklers	5.2	7.7	1196	8	3	747.5		1000	2200		3200	4.8 [a]
Cotton, 4 irrigations by sprinklers	5.2	7.7	1245	9	4	747.5		900	2200	900	4000	5.3 [a]
Cotton, drip	5.2	7.7	1329	10	5	747.5	400	600	1400	1200	4300	5.8 [a]
Tomato industry sprinkling	9.9	2.5	1561	21	9	73.1		1800	1700		5000	100.0
Peanut	2.9	3.2	1345	15	9	1013.2	400	2800	1200	600	5000	6.0
Wheat grain, rainfed only	7.1	6.0	615	2	1	299.0						4.0
Wheat grain, irrigated	7.1	6.0	689	3	1	299.0					2400	6.0
Wheat silage	7.1	6.0	681	3	1							30.0
Avocado	30.0	1.3	2835	35	26	2032.7	600	1500	1800	1800	9000	12.0
Wine grapes	10.6	1.2	5814	36	25	1328.9	800	1000	900		3100	20.0
Pecans	30.0	1.3	498	10	20	1528.2	1000		800		2500	2.5
Citrus	30.0	1.3	6644	35	9	573.1	900	900	1000	1000	9000	45.0

Monetary values are constant 1980 U.S. dollars

[a] Lint and grain yield

In addition to the sanitary damage resulting from effluent irrigation the salt concentration is higher in effluent than in freshwater. Soil salinity levels were calculated using a modification (Yaron et al., 1974, 1979) to a model proposed by Bresler (1967). The soil salinity level used in this study to calculate yield losses of crops is the average between the spring and the fall soil salinity, assuming that winter rains leached salt from the root zone. Yield losses are calculated according to coefficients proposed by Maas and Hoffman (1977) and Yaron et al. (1979); crop budgets are based on the work by the Israel Ministry of Agriculture (1980). Basic data for representative crops are reported in Table 11.2.

The farms differ in their land area, soil quality, irrigation technology, cropping pattern, freshwater quota, salt concentration of irrigation water, and distance from the suggested plant (Table 11.3). Farm A must participate in any coalition established because of its location.

Table11.3. Farms' major water and land characteristics

Participant	Total Irrigated Land Area, ha	Fruit Crops Included in Total, ha	Total Unirrigated Land, ha	Annual Freshwater Allotment, x 10^3m^3	Peak Months Water Quota, x 10^3m^3	Water per land Unit[a]		
						Irrigated Only m^3/ha	Total, m^3/ha	Peak Month, m^3/ha
Farm A	240.0	108.4	58.0	1600	159	6660	5360	660
Farm B	350.0	20.0	372.6	902	300	2580	1330	850
Farm C	196.9	42.5	91.1	850	138	4320	2930	700

[a] Without effluent

From the point of view of water use, the crops grown on the farms can be classified into four categories: (1) intensity-irrigated field crops, such as cotton (using alternative irrigation technologies, including drip irrigation), tomatoes, and corn for canning, (2) extensively-irrigated field crops, such as wheat (for grain), sorghum, and sunflower, (3) field crops not requiring irrigation, such as wheat (grain) and forage crops grown for hay or silage, and (4) perennial fruit crops, such as citrus, avocado, and vineyards. Some of these crops are sensitive to salinity (especially citrus and avocado). Detailed data on the technology applied in their growing and the estimates of their yield losses due to salinity can be found in Dinar (1984).

4.2. Results

The optimization model is first solved for the noncooperative conditions and offers the optimal solutions for each of the participants when they act independently. This is achieved by imposing $\delta_1 = \delta_2 = \delta_3 = 0$. Table 11.4

presents the results with respect to the treatment costs of the town, the income of the farms, and the shadow prices of freshwater.

Scrutiny of the shadow prices of freshwater in the non-cooperative situation (Table 11.4) suggests that the month of July is the most effective water constraint for each farm. The annual water constraint is effective for farms A and B, while the June water quota constrains farm A only. The high shadow prices of freshwater in July for all farms justify consideration of an additional water source to the region.

Table 11.4. Cost, gross income, and shadow prices of freshwater under the optimal noncooperative solution

	Town	Farm			Region's
		A	B	C	Total
Cost/income (10^3 US $)	-368	1940	1285	440	3297
		Shadow Prices of Freshwater, US $/m^3			
June quota		0.515	0	0	
July quota		1.074	0.559	1.161	
Annual quota		0.191	0.139	0	

Monetary values are constant 1980 US dollars

Regional cooperation in wastewater treatment for irrigation can arise among the town and some or all of the farms. The model solution for the region studied suggests that no cooperative agreement will be justified for a government subsidy of less than 15 percent of the overall treatment and the capital component of effluent conveying costs. If the subsidy is less than 15 percent, there is no incentive for any of the farms to use effluent in irrigation and no cooperative treatment plant will be set up. When the subsidy is 15 percent, cooperation between the town and farm A is justified, but farms B and C will be excluded. Only a 50 percent subsidy provides for full cooperation ("grand coalition") among all the potential participants.

A comparison of major results for cooperative situations, given 15 percent and 50 percent governmental subsidies, is shown in Table 11.5. It suggests that with a subsidy of 50 percent a plant of treatment level 2 (see Table 11.5) will be established. The 50 percent subsidy amounts to $497,000, and the regional income is increased by only $365,000. The environmental effects of such a subsidy are quite significant because the share of the regional wastewater used in irrigation (a good solution from the sanitary point of view) is 100 percent as compared with only 75 at a 15 percent subsidy level. At a 15 percent subsidy level only farm A uses effluent, while at 50 percent, all three farms participate. A comprehensive and conclusive discussion of the subsidy issue falls beyond the scope of this study. On the basis of (1) the positive environmental effects and (2) the fact that fresh-water is significantly subsidized too, a 50 percent subsidy was assumed for the continuation of the analysis.

Table 11.5. Comparison of major results in the regional optimization solution at different subsidy levels

Variables	15% Subsidy	50% Subsidy
Farms Participating	A only	all region's farms
Regional income[a], (10^3 US$)	3255	3622
Level of treatment	2	2
Wastewater to be treated, (10^3 m^3)	900	1200
Effluent used in irrigation [b], ($10^3 m^3$)	700	1015
Percent of regional wastewater	58	85
Percent of regional treated wastewater	7	100
Total treatment costs, (10^3 US$)	749	993
Subsidy, (10^3 US$)	113	496 [c]
Treatment costs to region, (10^3 US$)	636	487

Monetary values are constant 1980 US dollars

[a] Regional income with the subsidy included

[b] Gap between used effluent and treated wastewater is due to evaporation and infiltration in the storage site (model equations (2)-(3))

[c] Rounded values

Comparisons of other results for the noncooperative and cooperative situations are presented in Tables 11.6-11.8. The optimal cooperative solution enables each farm to efficiently reallocate the freshwater quota among the different land sections, to cultivate new land areas and to expand the irrigated crops by irrigating part of them with effluent.

The changes occurring due to the cooperative solution can be summarized as follows: (1) expansion of farm A's irrigated areas, no change in farm B's irrigated area and reduction in farm C's irrigated areas, (2) substitution of effluent for fresh-water by all producers; the quantity of freshwater use in the region is decreased by 330,000 m^3/yr, (3) increase in water input per land unit area; decrease in the area of unirrigated crops in the region, (4) expansion of certain crops and crops' irrigation procedures (new schedules of cotton irrigation, drip irrigation), and (5) reallocation of freshwater among each farm's various land sections according to the cropping patterns and the demand of sensitive crops for freshwater.

Table 11.6. Average treatment and conveying costs in a "Grand Coalition" cooperative setting

	Town	Farm A	Farm B	Farm C	Region
Cost/income in non-cooperative solution,[a] (10^3 US$)	-368	1940	1285	440	3297
Cost/income in cooperative solution,[a] (10^3 US$)	-497	2266	1365	488	3622
Overall treatment cost,[b] (10^3 US$)	993	...	...	...	
Subsidy for treatment, (10^3 US$)	497	...	...	...	
Treatment cost net of subsidy, (10^3 US$)	497	...	...	...	
Total effluent purchased, (10^3 m^3)	...	680	273	62	1015
Average treatment cost net of subsidy, (US$/$m^3$)	0.489	0.489	0.489	0.489	
Subsidy for transportation,[c] (10^3 US$)		25	33	3.8	
Transportation costs net of subsidy, (10^3 US$)	...	27	38	4	
Average transportation cost net of subsidy, (US$/ m^3)	...	0.039	0.143	0.068	
Overall average cost net of subsidy, (US$/$m^3$)	...	0.528	0.623	0.557	

Monetary values are constant 1980 US dollars
[a] Before redistribution of income
[b] Includes town's transport costs
[c] Only for the capital component

Tables 11.7 and 11.8 present the major changes induced by the cooperation for each farm: Farm A increases the irrigated area from 110 ha to 250 ha of which 170 are irrigated with effluent, substituting effluent for freshwater. Farm B does not increase its irrigated area but changes the cropping pattern by increasing the cotton area to 180 ha. Farm C decreases its irrigated area but increases the cotton area to 100 ha; farm C also equips 10 ha with drip irrigation for the cotton. A substantial decrease of 330,000 m^3 of freshwater in the region is another result of the cooperative solution (Table 11.8, row 1). This quantity remains at the disposition of the national system and can it be supplied to another region.

Results in Table 11.8 line 3 show that for some of the farms (B and C) there is an increase in the intensity rate of using water as is reflected by the ratio of total applied water per hectare. The high ratio for farms B and C explains the decrease or stability of their irrigated area. For farm A the ratio decreases because the irrigated area expands so much.

The relatively large source of irrigation water also enables the farms to transfer fresh water from land sections, which can be irrigated with effluent to land sections which are limited only to fresh water or that are being cropped with sensitive crops. These results are not presented.

Table 11.7. Land use and cropping patterns under the noncooperative and cooperative situations

	Farm							
	A		B		C		Region	
Row	Non-cooper-ative	Cooper-ative	Non-cooper-ative	Cooper-ative	Non-cooper-ative	Cooper-ative	Non-cooper-ative	Cooper-ative
1. Irrigated field crop area, ha	110.4	251.6	230.9	230.9	153.2	108.3	494.5	590.5
2. Irrigated fruit crops, ha	108.4	108.4	40.5	40.5	20.0	20.0	168.9	168.9
3. Total irrigated area, ha	218.8	360.0	271.4	271.4	173.2	128.3[a]	663.4	759.7
4. Effluent irrigated area,[b] ha	...	167.8	...	94.9	...	20.0	...	282.7
5. Percent of effluent irrigated,[c]	...	47.0	...	35.0	...	16.0	...	37
6. Unirrigated crop area, ha	183.5	113.3	424.3	424.3	...	59.0	607.8	596.6

[a] Including 100 ha newly equipped for irrigation
[b] Included in row 3
[c] 100 x rows 4/3

Table 11.8. Use of water under the noncooperative and cooperative situations

Row	Farm A Non-cooperative	Farm A Cooperative	Farm B Non-cooperative	Farm B Cooperative	Farm C Non-cooperative	Farm C Cooperative	Region Non-cooperative	Region Cooperative
1. Total fresh water use, x 10^3 m^3	1450	1280	760	700	500	400	2710	2380
2. Total effluent use, x 10^3 m^3	...	680	...	273	...	62	...	1015
3. Total applied per irrigated hectare, m^3	6600	5440	2800	3600	2900	3600	4100	4470
4. Total water use, x 10^3 m^3	1450	1960	760	973	500	462	2710	3395
5. Change in water use, %	+35		+28		-8		+25	
6. Use of freshwater in June, x 10^3 m^3	219	318	104	275	51	79	374	672
7. Use of effluent in June, x 10^3 m^3	...	210	...	34	...	20	...	264
8. Use of freshwater in July, x 10^3 m^3	241	209	183	182	90	118	514	509
9. Use of effluent in July, x 10^3 m^3	...	221	...	65	...	18	...	304

Values are in rounded numbers

5. SUMMARY

The paper presents a regional optimization model of municipal wastewater treatment and reuse in irrigation. Maximization of the regional income is constrained by the available production factors, given technologies of agricultural production, wastewater treatment technologies, prices, and environmental regulations.

The model was applied to a case study in a small region on the coastal plain of Israel. The empirical results show that without a subsidy there is no incentive for the farms in the region to use treated wastewater. Partial cooperation between the town and farm A is established when a subsidy level of 15 percent is given. In this case only 75 percent of the town's wastewater is treated in a treatment plant, while the remainder may cause environmental hazards. Comprehensive regional cooperation is possible only with a subsidy level of 50 percent. In this case all the town's wastewater is treated, and all the farms in the region use the effluent for irrigation. The town bears the increased treatment cost for all the other participants, while the farms increase their gross income and then compensate the town. This is discussed by Dinar et al. (1986).

The town and the farms in the region derive direct benefits from cooperation. The environment and the national water system, which are indirectly involved in the model also benefit from the cooperative solution. Environmental regulations are being followed in the cooperative solution. The total subsidy of US$559,000 to the region provides 330,000 m^3 freshwater to the national water system and the average of 1.69 US$/$m^3$ can be considered as a per cubic meter substitute for investment in new water resources. Shadow prices in Table 11.4 give a comparable range for this investment.

In accordance with the prevailing regulations in Israel the regional optimization model assumes that interfarm transfers of freshwater quotas are not permissible, and therefore redistribution of the additional regional income should be carried out only through monetary "side payments" by the farms to the town.

The acceptability of the regional cooperative solution depends on the establishment of a redistribution system, which is acceptable to all participants. Viewing the regional problem as a cooperative game with "side payments" allows the regional optimization problem and the income problem to be treated separately. This is an important feature of the problem and the model from the point of view of the computational burden. The problems of redistribution of income and specifically the payments to the town are treated in Dinar et al. (1986).

ACKNOWLEDGMENTS

This research was supported in part by grant I-101-79 from BARD, the United States-Israel Agricultural and Development Fund. Comments from J. Letey are most appreciated.

REFERENCES

Asano, T. and R. S. Mandancy. Water reclamation efforts in the United States in: "Water Reuse" (E. J. Middlebrooks, ed.). Ann Arbor Science. Ann Arbor, MI. 277-281. 1982.

Berezik, S. Nesher-Ramla reservoir planning report. Engineering and Consulting. Tel Aviv, Israel. April, 1982. (Hebrew)

Bresler, E. A model for tracing salt distribution in the soil profile and estimating the efficient combinations of water quality and quantity under varying field conditions. *Soil Science,* 104:227-233. 1967.

California State Water Resources Control Board. "Irrigation with Reclaimed Municipal Water: A Guidance Manual" (G. S. Pettygrove and T. Asano, eds.). Report 84-1 *WR*.. California State Water Resources Control Board, Sacramento. 1984.

Control Data Corporation. APEX-3 reference manual, version 1.1. Minneapolis, Minn. 1977.

Dinar, A. Economic analysis of regional wastewater treatment and use of effluent in irrigation and related cost benefit allocation schemes (PhD Thesis). Hebrew University of Jerusalem. Rehovot, Israel. 1984. (Hebrew)

Dinar, A., D. Yaron and Y. Kannai. Sharing regional cooperative gains from reusing effluent for irrigation. *Water Resources Research,* 22(2):339-344. 1986.

Dorfman, R. Conceptual model of a regional water quality authority in: "Models for Managing Regional Water Quality" (R. Dorfman, H. D. Jacoby, and H. A. Thomas, Jr., eds.). Harvard University Press. Cambridge, MA. 42-83. 1972.

Feigin, E., H. Bielorai, T. Kipniss and M. Gal. Use of municipal effluent for irrigating agricultural crops. *Special Publication,* 86. Agricultural Research Organization. Institute of Soil and Water. Bet Dagan, Israel. 1977. (Hebrew)

Giglio, R. J. and R. Wrightington. Methods for apportioning costs among participants in regional systems. *Water Resources Research,* 8(5):1133-1144. 1972.

Goodwin, H. L., Jr., R. F. Victurine and R. L. Lacewell. Economic implications of rural wastewater treatment alternatives. Paper presented at the Western Agricultural Economic Association Conference. San Diego, CA. 1984.

Israel Ministry of Agriculture. Extension Service. Computerized budgets for agricultural crops. Tel Aviv, Israel. 1980.

Loehman, E., J. Orlando, J. Tschirhart and A. Whinston. Cost allocation for a regional wastewater treatment system. *Water Resources Research,* 15(2):193-202, 1979.

Maas, E. V. and G. J. Hoffman. Crop salt tolerance-Current assessment. *Journal of Irrigation and Drainage Division American Society of Civil Engineers,* 103 (IR2):115-134. 1977.

Messer, J. International developments and trends in water reuse in: "Water Reuse" (E. J. Middlebrooks, ed). Ann Arbor Science. Ann Arbor, Mich. 549-576. 1982.

Moore, C. V., K. D. Olson and M. A. Marino. On-farm economics of reclaimed wastewater irrigation in: "Irrigation With Reclaimed Municipal Water: A Guidance Manual" (G. S. Pettygrove and T. Asano, eds.). *Report* 84-1 *WR,* California State Water Resources Control Board, Sacramento. 1984.

Nakamura, M. and E. D. Brill, Jr. Generation and evaluation or alternative plans for regional wastewater systems. An imputed value method. *Water Resources Research,* 15(4):750-756. 1979.

Papke, G. R., J. A. Smedile, and Greeley and Hansen Engineers. Wastewater treatment processes and cost estimating data. *Staff Paper,* 20. Northeast Illinois Planning Commission. Chicago, IL. April, 1977.

Rinaldi, S., R. Soncini-Sessa and A. R. Whinston. Stable taxation schemes in regional environmental management. *Journal of Environmental Economics and Management, 6*:29-50. 1979.

Tahal Israel Water Planning. Unrestricted use of effluent in agriculture. *Progress Report,* 2. Department of Effluent Reuse. Tel Aviv, Israel. 1978. (Hebrew)

Victurine, R. F., R. L. Lacewell and H. L. Goodwin, Jr. Economic implications of applying effluent for irrigation. Paper presented at the Western Agricultural Economic Association Conference. San Diego, CA. 1984.

Water Commission. Treated wastewater designated for irrigation. Advisory Committe for Establishing Rules on the Quality of Reclaimed Wastewater Used for Irrigation. Tel Aviv, Israel. 1978. (Hebrew)

Yaron, D., J. Shalhevet and E. Bresler. Economic evaluation of water salinity in irrigation. Report submitted to Resources for the Future. Hebrew University of Jerusalem. Rehovot, Israel. 1974.

Yaron, D., A. Dinar and S. Shamla. First estimates of prospective income losses due to increasing salinity of irrigation water in the south and Nagev regions of Israel. *Working Paper,* 7902. Center for Agricultural Economic Research. Rehovot, Israel. 1979. (Hebrew)

12

EVALUATING COOPERATIVE GAME THEORY IN WATER RESOURCES*

Ariel Dinar
University of California and USDA- ERS, Davis, USA

Aaron Ratner
Tel-Aviv College for Business Administration, Israel

Dan Yaron
Institute of Agricultural Economics, University of Oxford, United Kingdom

1. INTRODUCTION

Cooperative Game Theory (CGT) provides unique and efficient solutions in situations where decisions can be made both independently and collectively by a relatively small number of agents (players). For these cases, CGT is a better model of rational or efficient behavior than a market model. The latter model assumes a large number of players who do not interact except through market price. From this perspective CGT might be inferior to market models since the comparison among players is not fully captured on the basis of utility scales (as apposed to market price).

Game theory and CGT has been applied to a variety of situations where costs and benefits resulting from cooperation are allocated to the players (Littlechild and Thompson, 1977; Billera et al., 1970; Young et al., 1982; Selten, 1985; Young, 1985; Driessen, 1988; Kilgour et al., 1988). However, since researchers and professional journals generally prefer to report 'successes' in research rather than unsatisfactory results, the number of publications reporting the difficulties involved in applications is small (e.g., Heany and Dickin-

---

*Permission to publish this chapter was granted by Kluwer Academic Publishers. The chapter was originally published in *Theory and Decision,* 32:1–20, 1992.

son, 1982; Dinar et al., 1986; Williams, 1988). The impression obtained from the literature is apparently biased towards 'successes'. If this is true the motivation to work on issues related to the difficulties inherent to applications of CGT models might be reduced and the progress in the field impeded.

The purpose of this chapter is to point out some major difficulties, which arose in the authors' attempt to apply CGT concepts to two empirical problems. These difficulties tend to be general rather than specific to the problem studied, and therefore, seem worthy-of-exposure to a wider audience. The first study (Dinar, 1984; Dinar and Yaron, 1986) concerned a regional project designed for wastewater treatment and reuse as an irrigation water supply. It refers to a water scarcity problem faced by agricultural farms in the coastal plain region of Israel (and in principle, in many other irrigated areas). New water resources of lower quality, and higher cost can be developed if amount, quality and cost are sufficient and their allocation is acceptable. The second study (Ratner 1983; Yaron and Ratner, 1985, 1990) deals with regional interfarm cooperation in water use for irrigation and the determination of the optimal water quantity-quality (salinity) mix for each of the region's water users. In the first study a real problem and real players are presented; in the second study, only the farms considered are real and the study attempts to explore the potential for interfarm cooperation which does not yet exist.

2. CASE STUDIES

2.1. Regional Wastewater Treatment and Reuse for Irrigation

Water is a scarce resource in many parts of the world and expansion of the urban sector reduces the amount of water available for irrigation. Consideration of agricultural reuse of treated municipal wastewater can solve the dual problem of providing additional and substantial source of irrigation water on the one hand, and reduce possible environmental pollution on the other hand.

A regional approach to municipal wastewater management through treatment and irrigation reuse (subject to public health regulation) was analyzed by Dinar and Yaron (1986). This study showed that a region considering such an endeavor faces the following interrelated problems: 1) Determination of the regional boundaries with due consideration of treatment plant capacity and the capacity and layout of the wastewater and effluent conveyance systems; 2) Determination of the wastewater treatment level; 3) Allocation of the effluent (treated wastewater) to the agricultural farms within the region; 4) The selection of optimal cropping patterns; 5) Cost allocation to the participants; and 6) Level of government subsidy, if needed.

A non-linear-separable mixed integer mathematical programming model, which includes these components, was formulated, designed to analyze the economic potential for regional cooperation in the treatment of municipal wastewater and reuse for irrigation. An efficient regional solution, which maximizes the region's income under cooperation, was provided by the model. As a result of the cooperation the farmers may increase their irrigated acreage and benefits; the treatment cost can rise based on higher treatment levels than those required by health standards used for direct disposal. The joint treatment cost could be borne by the farmers, by the municipal sector and/or by a government subsidy, if needed. (The subsidy possibly being justified by an improved environmental quality.)

In order to correspond with the prevailing institutional arrangements, it was assumed that freshwater from the individual farm's quotas is not transferable and the only way to compensate the participants (if necessary) is through direct income transfers (side payments). Thus, the problem falls within the category of a transferable utility situation with an important property: the possibility of first finding the efficient solution (maximal regional income), and at a second stage considering the redistribution of income among the participants. Such a possibility reduces the conceptual and computational difficulties.

2.1.1. *The Nature of the Cost Allocation Problem*

The region referred to in this study consists of three farms (*A, B, C*), a town, and a potential treatment plant. The regional 'game' allocates the additional costs and benefits. The situation in the region indicates that: (1) no cooperation can occur without the participation of the town; and (2) farm *A* must be included in any cooperative arrangement because of its location in the region. The players in the game, the three farms and the town, will be referred to as players 1, 2, 3, and 4, respectively.

Let N be the set of all feasible coalitions in the game; s^* is a feasible coalition if and only if $s^* \in N$ and $1, 2 \in s^*$. The non-cooperative coalitions are $\{i\}$, $i = 1, 2, 3, 4$, and the grand coalition is N. The number of *a priori* feasible coalitions in the region is less than the potential number of coalition arrangements ($2^4 - 1$):

Non cooperative coalitions	$\{1\}, \{2\}, \{3\}, \{4\}$
Partial coalitions	$\{1, 2\}, \{1, 2, 3\}, \{1, 2, 4\}$
Grand coalition	$\{1, 2, 3, 4\}$

The regional model was applied to any feasible coalition s^* in the region with the objective function being the maximization of the aggregate region's income (including players that act independently). For example, in the case where the partial coalition $\{1, 2, 3\}$ exist, the model includes a joint objective function for players 1, 2, and 3, and a separate objective function for player 4 who acts independently. The regional income is the aggregation of the partial coalition's objective function value and the value of the objective function of player 4. The model was structured to also determine whether coalition s^* will be created or not. Assuming that coalition $s^* - i$ is economically viable, coalition s^* will be created only if $V(s^*) \geq V(s^* - I) + V(i)$, with $V(s^*)$ being the income of coalition s^*. Otherwise, results for the optimization model would suggest the existence of two coalitions: $s^* - i$ and i, with player i acting independently (for more details see Dinar and Yaron, 1986). Here, the emphasis is on allocation of the benefits derived from regional cooperation. Recall that because side payments are possible (transferable utility), the regional optimization and the cost-income allocation problems can be solved in two subsequent and disjoint stages. This is not the case when utility is not transferable; the optimization problem is then constrained by income distribution considerations.

Table 12.1. Income of players in the optimal regional solutions for different coalition combinations, the value for the different coalitions, and the incremental income values before redistribution[a] (case study I)

	Income of player						
	1	2	3	4	Value for Coalition s^*	$\sum_{i \in s^*} V(i)$	$V(s^*) - \sum_{i \in s^*} V(i)$
Coalition	Town	Farm *A*	Farm *B*	Farm *C*	$V(s^*)$		
	(1)	(2)	(3)	(4)	(5)	(6)	(7)
{1}	-368				-368	-368	0
{2}		1940			1940	1940	0
{3}			1285		1285	1285	0
{4}				440	440	440	0
{1, 2}	-410	2267			1857	1571	285
{1, 2, 3}	-498	2275	1370		3147	2857	290
{1, 2, 4}	-395	2205		488	2299	2011	297
{1, 2, 3, 4}	-497	2266	1365	488	3622	3297	325

Rounded figures

[a] Column (1) shows the (treatment) cost borne by the town before compensation; Columns (2)-(4) show the income generated for Farms *A, B*, and C before payment for the treated water. Column (5) is the summation of columns (1) through (4); Column (6) is the summation of the s^*'s players' income while they are acting alone. Column (7) is the difference between (6) and (5), representing the normalized values of the game.

Table 12.1 presents the results of the optimization problems for the eight possible coalition arrangements. Values in that table suggest that the incremental income of a coalition generated by players 1 and 2 is US$285,000. This is a result of an increase in the gross income generated by player 2 from US$1,940,000 to US$2,209,000, and a very minor increase in the treatment costs of US$41,500. If players 3 and 4 join coalition {1, 2} individually, it increases the income by a negligible amount[1] (in US$000). When they both join coalition {1, 2}, the additional net incremental income (in normalized values) is only US$40,000 (325,000-285,000). This implies that farm *A* is the significant contributor to the regional income, while farm *B* and *C* fall far behind. Common sense suggests that this fact should be considered in any income allocation scheme.

2.1.2. *The Cost Allocation Schemes*

In this section the results for several game theory allocation schemes are presented and discussed. The first scheme that has been used in many empirical studies is the Core. In the game considered here, the core is a four dimensional polyhedron. Its extreme points (computed following Shapley, 1971) are presented in Table 12.2. Scrutiny of the data in this table shows that the core is quite large; it suggests only bounds for the allocations to the players and may serve as a guideline for arbitration or while referring to other allocation criteria.

Table 12.2. Extreme points of the core in the regional game (case study I)

Maximum profit allocation to player i (US\$000)[a]				Coalition formation sequences leading to this allocation
1	2	3	4	
0	285	5	35	1234
0	285	38	2	1243
0	290	0	35	1324 3124
0	325	0	0	1342 1432 3142 3412 4132 4312
0	287	5	0	1423 4123
285	0	5	35	2134
285	0	38	2	2143
290	0	0	35	2314 3214
325	0	0	0	2341 2431 3241 3421 4231 4321
287	0	38	0	2413 4213

[a] Rounded figures

Results in Table 12.3 are for three uniquely determined solutions—the Nucleolus (Schmeidler, 1969), computed as an extension to the determinate core (and the determinate Least Core), the Shapley Value (SV) (Shapley, 1953), and the Generalized Shapley Value (GSV) (Loehman and Winston, 1976) which is an extension to the original SV in the sense that it refers only to coalitions which can realistically be formed rather than to all permutationally conceivable coalitions. These solutions were derived on the basis of: (1) prior sound and acceptable axioms, and (2) the assumption that utility is linear in money when dealing with incremental income or the cost of a large economic unit, with the increment being small relative to the total income or cost. These questions follow: Can these solutions be justified in terms of common sense? Can they be regarded as acceptable?

There are at least two difficulties inherent in these solutions. The first difficulty is the allocations to the town and to Farm *A,* which are equal in all three schemes. This reflects their symmetry as players and their bargaining power. However, the increase in income of Farm *A* ranges from 7 percent to 8 percent, while the decrease in the expenses of the town is in the range of 39 percent to 42 percent compared with the non-cooperative situation. The relative additional income to Farm *A* is very small, while the relative reduction in the costs borne by the town is significant. Moreover, Farm *A*, which uses effluent for agricultural production is exposed to a considerably higher risk level than the town which is in charge of a simple treatment process. One could argue, therefore, that a fair allocation should provide a risk premium to Farm *A* and should discriminate in favor of Farm *A* and against the town, because they are not, symmetric players in the sense of risk. The conclusion from the above discussion is that the simplistic assumption of utility as linear in money is not proper with reference to the problem analyzed, and should be modified.

Table 12.3. Distribution of the incremental income among participants in the regional cooperation according to alternative allocation schemes (case study I)

Allocation scheme	Player			
	1 Town	2 Farm *A*	3 Farm *B*	4 Farm *C*
Non-cooperation income (US$000)	-368[a]	1940	1285	440
Nucleolus				
Allocation (US$000)	149.5	149.5	13	13
% of regional incremental income	46	46	4	4
% of regional income with no cooperation	41[b]	8	1	3
Shapley				
Allocation (US$000)	153	153	10	9
% of regional incremental income	47	47	3	3
% of regional income with no cooperation	42[b]	8	1	2
Generalized Shapley				
Allocation (US$000)	142.7	142.7	21.4	21.4
% of regional incremental income	44	44	7	6
% of regional income with no cooperation	39[b]	7	2	1

[a] Represents the expenses decrement rate
[b] Rounded figures

The second difficulty is related to the coalitions referred to and the probabilities of their formation. The difference between the allocations to Farms *B* and *C* according to the SV and the GSV solutions clearly illustrate this point. While not so visible, the same difficulty is pertinent to the core solution and its derivatives (the Least Core and the Nucleolus).

2.2. Optional Regional Water Quantity-Quality Mix for Agricultural Users

In many parts of the world the allocation of irrigation water to users is based on water rights (quotas), which have been institutionally determined in the past. Generally, these rights have not been modified to address changes in farming systems and new technologies. The inevitable result is inefficiency in the interfarm and interregional allocation of water (Frederick and Gibbons,

1985). The inefficiency of the institutional water allocation system has been exacerbated recently by the increased use of low quality water (e.g. brackish water, reuse of drainage water) for irrigation in regions, which suffer from water scarcity. For example, Vaux and Howitt (1984) have analyzed the efficiency of possible institutions for water transfers among regions under conditions of relative water scarcity in California.

One possible way to increase the efficiency of water allocation among farms within regions is through the voluntary establishment of regional water associations or farmer cooperatives. The members of the associations will be able to exchange water quotas among themselves, and with other entities (e.g., Bureau of Reclamation, state water agencies). It should be noted that such arrangements already exist and operate in certain regions with irrigated agriculture (Wahl, 1989).

2.2.1. *The Regional Problem*

Consider a region with I farms and a given regional water allotment (**G**) of a given quality (salinity) level (R_0). **G** is the sum of the individual farm quotas (G_i),

$$\mathbf{G} = \sum_i G_i \,.$$

Assume that the water authority can supply the region with a larger quantity of water at the cost of increasing its salinity.[2] For a given $\mathbf{G}$, the substitution between water quantity ($\mathbf{B}$) and its salinity (R) is subject to a substitution function determined by the water authority.

$$F\left(\mathbf{B}, R \middle| \mathbf{G}\right) = 0 \,. \tag{1}$$

It is further assumed that R must be the same for all users in the region. Any decision regarding R and the quantity of water with this salinity level must be mutually agreed upon by *all* the farms in the region.[3] The prospect for additional income provides the essential incentive for regional farms to cooperate within the framework of the water users' association (e.g., a water district). Farm i's income y_i is a function of B_i and R, with B_i denoting the quantity of water with quality R allocated to farm i:

$$y_i = y_i\left(B_i, R\right) \qquad i = 1,\ 2,\ \ldots.,\ I \tag{2}$$

Note (obviously) that

$$\mathbf{G} = \sum_i B_i = \mathbf{B} \,.$$

The cooperatives problem is to determine simultaneously: (1) the optimal quantity-salinity (**B** – R) mix for the region, and (2) the quantity of water (B_i) of salinity $R > R_0$ allotted to each farm. The optimality conditions for the relationship between **B** and R are derived from the following problem:

$$\text{(3)} \qquad \max W = \sum_i \lambda_i y_i$$

subject to:

$$\text{(4)} \qquad F(\mathbf{B}, R, \mathbf{G}) = 0$$

$$\text{(5)} \qquad \sum_i B_i - \mathbf{B} = 0$$

with W being the cooperative's welfare function and λ_i the relative weight assigned to the ith farm income ($\lambda_i \geq 0,\ \Sigma_i \lambda_i = 1$). By parametrically varying the λ_i weights, the efficiency frontier in the I farms' income space can be derived. The efficiency frontier is the locus of Pareto-optimal points with the characteristic that any move from such a point to improve the income of one farm must necessarily reduce the income of some other farm(s).

2.2.2. *A Quasi-empirical Application*

Ratner (1983) and Yaron and Ratner (1985, 1990) have applied a linear programming model to the empirical estimation of an income efficiency frontier presented earlier for a regional cooperative in Israel. We first present partial results for a case of two farms only. The advantage of considering a two-farm case is that it can be graphically depicted; its disadvantage is that it does not illuminate issues involving cooperation of three or more.

The following assumptions underlie the analysis: (1) The regional supply substitution relationship between water salinity R and quantity **B** is determined by the water authority and is presented to the region, which can change the prevailing **B** R mix. (2) The **B** R combination for the region is determined as a part of the cooperative agreement within the region as are the quantities of water allocated to each farm. The same salinity level R of water is supplied to all regions' farms. (3) The income of each cooperating farm must be equal to or higher than its income before cooperation (individual rationality). (4.1) The cooperation involves both water and direct-income transfers (transferable utility situation, hereafter referred to as *TU*); or: (4.2) The cooperation is restricted to transfers of water only with the distribution of income among the farms determined by those transfers. For example, Farm 1 transfers water to Farm 2 in June, and a reverse transfer takes place, say in July. This is the non-transferable utility situation, hereafter referred to as *NTU.* Direct transfers of income (money) as side payments are possible.

Assumptions 1 through 4.1 or 1 through 4.2 will be referred to as scenarios 3 and 4, respectively. A few other scenarios that were analyzed (Ratner, 1983) will not be discussed here.

The results for scenarios 3 and 4 are presented in Figure 12.1, which portrays the transformation curves of income between the two farms. The point *Q* represents the income derived in the absence of cooperation. The feasible region for income allocation lies 'northeast' of *Q.* The point *N* corresponds to the maximal cooperative income under scenario 3 but it lies outside the feasible zone; the feasible part of the income transformation curve is *MS.* For scenario 4 the income transformation curve is *EF.* Due to cooperation the regional income rises by 22.5 percent under scenario 3 (*TU*) compared to the non-cooperative regional income. Under scenario 4 (*NTU*) the regional income rises by 5.9 to 14.5 percent compared to the non-cooperative regional income, depending on the particular point chosen on *EF.*

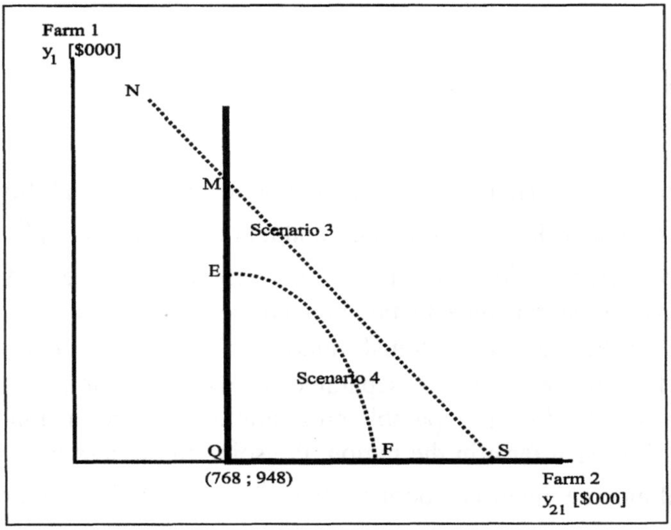

Figure 12.1. Income transformation curves between Farm 1 and 2 (scenarios 3 an 4)

Clearly, the potential for additional income due to cooperation is higher when side payments are possible. However, the soundness of such transfers with no *a priori* reference to the price per unit of water may be questioned, especially considering the general resentment of farmers to adopt side payments as a policy .On the other hand, if the transfer of lumpsum payments will satisfy some notion of fairness, the objection may be mild or even not exist, given that the institutional objection to rent benefits resulting from water quotas is removed. This issue is unresolved and remains open for further study.

2.2.3. *Three Farms' Cooperation -NTU Situation (Scenario 4)*

In this case, regional income maximization is subject to income distribution constraints; both are linked and should be solved simultaneously. Under these conditions, except for the Nash-Harsanyi solution (Nash, 1950; Harsanyi, 1959), income allocation solutions are subject to computational and, in some cases, conceptual difficulties.

3. THE NASH-HARSANYI SOLUTION

The Nash-Harsanyi solution can be derived using the following model:

$$\text{(6)} \qquad \max\ Z = \prod_i \left(y_i^R - y_i^0\right)$$

subject to

$$\text{(7)} \qquad y_i^R \geq y_i^0 \qquad i = 1,\ 2,\ 3$$

and technological restrictions including the demand and availability of limited input factors such as land, water, labor, and marketing factors. y_i^R and y_i^0 are the i th farm income with and without cooperation respectively. As assumed, water quotas can be transferred among the cooperating farms.

Let $\mathbf{Z}$ be the logarithmic transformation of Z ($\mathbf{Z}$ is a monotonic function of Z) in (6). $\mathbf{Z}$ is now a separable objective function that allows the problem to be solved using separable programming routines available on most computers. For each farm in the region, the solution process involves finding the optimal income under no cooperation (y_i^0) and then finding a regional solution according to (6) and (7) above. This is computationally simple compared with the other game theory models applied in this study.

The optimal water quality R is determined by solving the problem parametrically, for several discrete chosen levels of R (Table 12.4). Scrutiny of this table suggests that the distribution of total income generated among the three cooperating farms, in percentage terms, is about the same as before the cooperation was established. From this point of view, the results seem intuitively acceptable. The aggregate income in the case of cooperation rises by 8 percent compared with the non-cooperative situation, while the incomes of farms 1, 2, and 3 rise by 9, 5, and 11 percent, respectively.

Table 12.4. Nash-Harsanyi solution for a three-farm cooperative with reference to scenario 4 (case study II)

	Farm			
	1	2	3	Total
Income with no cooperation	946	769	605	2320
% of regional income	41	33	26	100
Income generated on farm in the cooperative solution	1031	811	672	2514
% of regional income	41	32	27	100
Incremental income	85	42	67	194
% of regional incremental income	43	22	35	100

Rounded figures

4. THE CORE SOLUTION

The Core of a game is formally defined as the set of feasible allocations which can be improved upon by no coalition *s* in *S*, where *S* denotes the set of all possible coalitions (Hildenbrand and Kirman, 1976). In the context of the empirical problem referred to, the question arises as to which coalitions are indeed feasible and realistic. In some cases, an objective assessment of their technical feasibility can be derived from the design of the regional water system. Regarding their socio-economic feasibility, difficulties may arise in view of possible strategic behavior by the farms. Suppose (contrary to the assumption previously made) that all pair-wise coalitions are feasible.

The Core of the three-farm cooperative in the *NTU* situation is then defined by the following relationships:

(8.1) $y_2 \geq g_{21}(y_1)$

(8.2) $y_2 \geq g_{23}(y_3)$

(8.3) $y_3 \geq h_{31}(y_1)$

(9) $y_1 + y_2 + y_3 = V(1,\ 2,\ 3)$

where $g_{2i}(y_i)$ is the income transformation function for the partial coalition of Farms 2 and *i* (*i* = 1, 3); $h_{31}(y_1)$ is the income transformation function for Farms 3 and 1, and *V*(1, 2, 3) is the aggregate income of the three-farms cooperative (Grand Coalition). Note that its magnitude depends on the income allocations y_i. Note also that any point $[g_1(y_1), y_2]$ satisfying $y_2 \geq g_{21}(y_1)$

satisfies also $y_1 \geq g_{21}^{-1}(y_2)$; similar inverse relationships hold for $g_{23}(y_3)$ and $h_{31}(y_1)$; restrictions (8.1)-(8.3) satisfy the individual rationality requirements of the farms as well.

The Core of the three farms cooperative game was parametrically solved by maximizing y_i^R subject to restrictions (7) and (8.1)-(8.3) for various combinations of y_2 and y_3 (arbitrarily chosen), and subject to technological restrictions concerning the availability of limited input factors. The non-linear functions $g_{21}(y_1)$, $g_{23}(y_3)$and $h_{31}(y_1)$ were piece-wise linearly approximated and a separable programming routine was used. The problem was separately solved for the different pre-selected values of R. It is noted that the optimal R may differ from one region (or point) to another on the Core surface. The computed Core is presented in Figure 12.2 in terms of the additional income generated by cooperation. The point (0, 0, 0) represents the income obtained by the three farms with no cooperation, and the values along the three axes are expressed in terms of the additional income (normalized game). The additional incomes of Farms 2 and 3 are clearly marked in the graph, while the income of Farm 1 is represented by the height of the tri-dimensional surface (the Core). The shadowed area in the $y_2^* y_3^*$ plane represents combinations of $y_2^* y_3^*$; which do not satisfy the group rationality conditions (8.1)-(8.3).

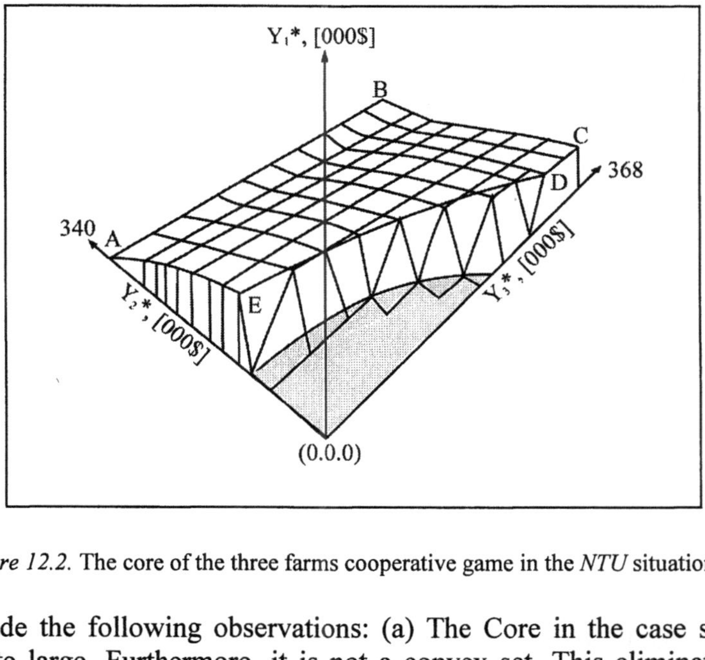

Figure 12.2. The core of the three farms cooperative game in the *NTU* situation

We made the following observations: (a) The Core in the case studied here is quite large. Furthermore, it is not a convex set. This eliminates the possibility of generating new points in the Core by convex combinations of selected points lying in the Core. (b) Heavy computational burden is involved in computing the Core for the three farms cooperative game. The computations might be prohibitive for cooperatives (coalitions) with larger number of members.

In view of the above, one may wonder whether the Core is a practical concept in applications such as the one dealt with here.

5. CONCLUSION

The paper reviews two case studies involving cooperation in the use of water for irrigation under conditions characterized by a general trend of increasing water scarcity and salinity. The focus of the paper, the application of cooperative game theory models to income distribution problems associated with the income maximization of regional individuals or institutions, is a difficult task—even if it is done with full cooperation for their own use (e.g., Keeney and Raiffa, 1976). The difficulties might be significantly exacerbated if the assessment of the utility functions refers to entities involved in a dispute regarding allocation of gains from cooperation. Thus, in situations where the assumption of linear utility in money is not acceptable, a two-stage analysis is required: (1) assessment of the utility functions of the players, and (2) cost/benefit allocation among the players. This seems to be quite cumbersome.

Another problem relates to the feasibility and probability of forming partial coalitions. This clearly affects the allocations to the individual players in the *TU* situation, in the Shapley and Generalized Shapley Values, and also affects the Core and the solutions derived from it. In a comprehensive comparison of several CGT solution concepts performances (allocation of 14 project costs and benefits among different project purposes), Williams (1988) concludes that the SV received weaker empirical support than other solution concepts discussed there, including cases where the SV lies outside the Core.

The third major difficulty arose in the second case study (*NTU* situation). The computation of the Core was very cumbersome and with the Core being large and a non-convex set, the usefulness of the whole procedure could be questioned. On the other hand the Nash-Harsanyi solution in the *NTU* situation, in the second case study, yielded results which might be acceptable. Furthermore, the burden of the computational procedure seems reasonable (see also Williams, 1988).

The application of the Nash-Harsanyi model to the problem discussed here (and other problems) seems promising since: a) the underlying axioms are relatively simple and easy to accept, and b) it is computationally simple and relatively easy to manage. One difficulty envisaged in its application to real life can be understood by the link between the underlying axioms and the nature of the solution. This can be solved practically by cooperatives. The first case study is concerned with municipal wastewater treatment and sharing the gains to the regional cooperative resulting from the reuse of effluent for irrigation. The emphasis is on the allocation of wastewater treatment costs among the town, which is in charge of the treatment plant, and the farms, which use the effluent for irrigation. The cost allocation problem falls within the category of transferable utility and therefore, can be treated independent of the regional income maximization problem, after this has been solved. This procedure has considerable computational advantages. The second case study involves farms operating in a region where fresh water is scarce and additional supplies have higher salinity. Cooperation occurs in determining regional water quality (salinity) and exchange of water quotas among the farms. The cooperation within the region and the related income distribution problems are interrelated and must be solved jointly.

The difficulties in applying game theory models to the derivation of cost/income allocation schemes for the case studies described above are summarized here. The general question of interpersonal utility comparison was widely discussed in the context of game theory (Shapley, 1988; Young, 1988) but no clear solutions to the concerns raised in the current paper were provided. Indeed one could include external considerations and preferences, such as weights on the different players to reflect the way their utilities should be evaluated which of course could generate considerable opposition (see, for example, Rawls, 1971). Kalai and Samet (1980) discuss this topic and suggest a series of formulas to calculate a weighted SV. The basis for the weights is the relative size of the group represented by each player. For example, in a game involving several communities, the population in each community serves in establishing the weights. This of course ignores the type of problems discussed above.

The simplifying assumption of utility as linear in money that was applied to all players, leads to questionable results in the first case study. The assessment of utility functions for involving an authoritative arbitrator may use the model as an aid in the arbitration procedure.

To conclude, in the two case studies reviewed, each addressing a real life problem of considerable concern to farmers and policy makers, game theory approaches did not provide indisputable solutions to the cost/income allocation problems. The applications of game theory model provided better understanding of the problems and could be useful in bargaining and arbitration (for an extensive discussion, see Binmore and Dasgupta, 1989) toward deriving of the agreeable solutions, but no clear-cut solutions were obtained. The applications described also pointed to some conceptual difficulties which appear to be inherent to the models applied. The challenge of being able in the future to design objective cost/income allocation schemes or at least to limit the scope for arbitration is great. The ultimate goal of this paper is to suggest that considerable work is needed by both model developers and model users in order to proceed in this challenging direction.

Difficulties arising in empirical application of cooperative game theory models should be reported and explored. There is no better way to conclude this paper than to quote Shubik (1985):

> "As is not uncommon with other theories, the current growth and success of game theory have led to a deeper understanding of its weaknesses and of many of the weaknesses in economic theory. In particular, the need for further development and experimentation with solution theories is clear."

ACKNOWLEDGEMENTS

The research leading to this paper was supported in part by a grant I-101-79 from BARD, The United States-Israel Agricultural Research and Development Fund. Yakar Kannai and G. T. Jones provided useful comments.

REFERENCES

Billera. L. J., D. C. Health and J. Raana. Internal telephone billing rates—a novel application of non-atomic game theory. *Operations Research 26.* November December, 1978. 1970.

Binmore, K. and P. Dasgupta. "Economic Organizations as Games." Basil Blackwell, New York. 1989.

Dinar, A. Economic analysis of regional wastewater treatment and use of effluent in irrigation and related cost benefit allocation schemes (PhD Thesis). The Hebrew University of Jerusalem. Rehovot, Israel. 1984. (Hebrew)

Dinar, A. and D. Yaron. Treatment optimization of municipal wastewater and reuse for regional irrigation. *Water Resources Research,* 22:331-338. 1986

Dinar, A., D. Yaron, D. and Y. Kannai. Sharing regional cooperative gains from reusing effluent for irrigation. *Water Resources Research,* 22:339-344. 1986.

Driessen, T. "Cooperative Games, Solutions and Applications." Theory and Decision Library, Series C: Game Theory, Mathematical Programming and Mathematical Economics. Kluwer Academic Publishers, MA. 1988.

Harsanyi, J. C. A bargaining model for the cooperative n-person game in: "Contributions to the Theory of Games" 1-4, A (W. Tucker and R. D. Luce, eds.). Princeton University Press, NJ. 325-355. 1959.

Heany, J. P. and R. E. Dickinson. Methods for apportioning the cost of a water resources project. *Water Resources Research,* 18:467-482. 1982.

Hildenbrand, W. and A. P. Kirman. "Introduction to Equilibrium Analysis." North-Holland, New York. 1976.

Frederick, K. D. and D. C. Gibbons (eds.). "Scarce Water and Institutional Change." Resources for the Future. Washington, DC. 1985.

Kalai. E. and D. Samet. Weighted Shapley values in: "The Shapley Value-Essays in Honor of Lloyd S. Shapley" (A. E. Roth, ed.). Cambridge University Press, New York. 83-99. 1988.

Keeney. R. L. and H. Raiffa. "Decisions with Multiple Objectives." Wiley and Sons. New York. 1976.

Kilgour. D. M., N. Okada, A. and Nishikori. Load control regulation of water pollution: an analysis using game theory. *Journal of Environmental Management,* 27:179-194. 1988.

Littlechild, S. C. and G. D. Thompson. Aircraft landing fees: a game theory approach. *The Bell Journal of Economics and Management Science,* 8: 186-204. 1977.

Loehman, E. T. and A. Winston. A generalized cost allocation scheme in: "Theory and Measurement of Economic Externalities" (A. Stevens and Y. Lin, eds.). Academic Press, New York. 87-101. 1976.

Nash, J. F. The bargaining problem. *Econometrica,* 18:155-162. 1950.

Ratner, A. Economic evaluation of regional cooperation in water use for irrigation-optimal allocation of water quantity and quality and the related income distribution (unpublished MsC Thesis). The Hebrew University of Jerusalem. Rehovot, Israel. 1983. (Hebrew)

Rawls. J. "A Theory of Justice." Harvard University Press, MA. 1971.

Schmeidler, D. The nucleolus of a characteristic function game. *SIAM, Journal of Applied Mathematics,* 17(6):1163-1170. 1969.

Shapley, L. S. A value for n-person games in: "Annals of Mathematics Studies, Contribution to the Theory of Games" (H. W. Kuhn and A. W. Tucker, eds.). 11(28):307-318. 1953.

Shapley, L. S. Cores of convex games. *International Journal of Game Theory,* 1:11-26. 1971.

Shapley, L. S. Utility comparison and the theory of games in: "The Shapley Value-Essays in Honor of Lloyd S. Shapley" (A. E. Roth, ed.). Cambridge University Press, NY. 307-319. 1988.

Shubik, M. What is game theory trying to accomplish? – Comment in: "Frontiers of Economics" (K. J. Arrow and S. Honkapohja, eds.). Basil Blackwell, NY. 88-97. 1985.

Selten, R. What is game theory trying to accomplish? – Comment in: "Frontiers of Economics" (K. J. Arrow and S. Honkapohja, eds.). Basil Blackwell, NY. 77-87. 1985.

Vaux, H. J. and R. W. Howitt. Managing water Scarcity: an evaluation of interregional transfers. *Water Resources Research,* 20(7):785-792. 1984.

Wahl, R. W. "Markers for Federal Water." Resources for the Future. Washington, DC. 1989.

Williams, M.A. An empirical test of cooperative game solution concepts. *Behavioral Science,* 3:224-230. 1988.

Yaron. D. and A. Ratner. Efficiency and game theory analysis of income distribution in the use of irrigation water. Report Series 14. Institute of Agricultural Economics, University of Oxford. 1985.

Yaron, D. and A. Ratner. Regional cooperation in the use of irrigation water: efficiency and income distribution. *Agricultural Economics,* 4:45-58. 1990.

Young, H. P., N. Okada and T. M. Hashimoto. Cost allocation in water resources development. *Water Resources Research,* 18(1):463-475. 1982.

Young, H. P. "Cost Allocation." Eisevier Scientific Publishers, NY. 1985.

Young, H. P. Individual contribution and just compensation in: Shapley Value-Essays in Honor of Lloyd S. Shapley" (Roth, A. E., ed.). Cambridge University Press, NY. 267-278. 1988.

---

[1] For example, if player 3 joins {1, 2} the incremental income (US$000) is:

$$\begin{aligned} V(1,\ 2,\ 3) - V(1,\ 2) &= [3147 - (-368 + 1940 + 1285)] - [1857 - (368 + 1940)] \\ &= 290 - 285 = 5 \end{aligned}$$

[2] There are several technically possible choices for this, such as providing brackish water or treated wastewater, which are generally more saline than fresh water.

[3] Alternatively it can be assumed that *all farms in any part* of the region must have the same salinity level in the water supply.

13

ADOPTION AND ABANDONMENT OF IRRIGATION TECHNOLOGIES*

Ariel Dinar
University of California and USDA- ERS, Davis, USA

Dan Yaron
The Hebrew University of Jerusalem, Rehovot, Israel

1. INTRODUCTION

The diffusion of innovations has long been a major topic in the context of technological change. Most of the empirical economic studies on diffusion of technologies have estimated rates of adoption and levels of adoption until the stage when the process reaches its ceiling (e.g. Jarvis, 1981; Jansen, Walker et al., 1990). Less attention has been devoted to the stage when the innovation is abandoned, which occurs with the same frequency in the history of the technology. A discontinuance of technology can be the result of technological substitution (Fisher and Pry, 1971; Rogers, 1983; Cameron and Metcalfe, 1987), which creates technology cycles.

In the development of new technologies it is essential for the developer, or the policy maker, to estimate the expected life span of the technology in order to analyze the effects of possible policy variables on the resulting number of users of that technology (or any other measure for intensity of use). Several studies have recognized the importance of irrigation technologies in the process of agricultural development (Hayami and Ruttan, 1971; Kulshreshtha, 1989). The economic literature on irrigation technology diffusion has generally provided information on diffusion of one or at most two technologies (Fishelson and Rymon 1989; Casterline et al., 1989). Less attention has been devoted to the abandonment phase of the technologies. This is

*Permission to publish the chapter was granted by the Elsevier Science Publishers. The chapter was originally published in *Agricultural Economics,* 6:315–332, 1992.

probably due to a lack of information on use of various technologies over time.

The purpose of this chapter is to extend the existing literature by depicting and estimating the diffusion-abandonment process of several irrigation technologies. The next section provides a conceptual framework for the analysis, which includes both the procedure for estimating technology cycles and the framework for estimating diffusion-abandonment and diffusion curves for several technologies. This framework is applied to survey data from citrus groves in Israel. As compared with previous studies, this database addresses several irrigation technologies. The estimates for the diffusion curves are used to demonstrate and estimate policy effects on the diffusion of irrigation technologies.

2. CONCEPTUAL FRAMEWORK

The introduction of any technology can be described as composed of two phases: in the first phase, the technology is introduced to an increased number of users (or any measure of use, such as acres or hectares). This phase is generally defined in the literature as the diffusion of the technology. The second phase is characterized by declining use of that technology. The economic literature has concentrated mainly on estimating diffusion curves for technology (e.g. Griliches, 1960; Jarvis, 1981). However, it is also important for policymakers to know the rate and time at which a technology will be abandoned. The analysis provided in this section distinguishes (using the terminology suggested earlier) between two groups of technologies: (1) technologies that are in a process of abandonment or have already been abandoned, and (2) technologies that are still in the diffusion process.
For the first group of technologies, a procedure is suggested to estimate the technology cycle and is applied to data for citrus groves. The technology cycle provides data used to estimate a quadratic expression of share of users for a given technology over time (a diffusion-abandonment pattern), which is applied to estimate the time of a complete discontinuance of that technology. For the second group of technologies diffusion logistic curves are estimated. Crop yield price, water price, and government subsidy for irrigation equipment are used to explain diffusion rates.

2.1. A Procedure for Estimating Technology Cycles

The term 'innovation cycle' in agriculture was used by Kislev and Shchori-Bachrach (1973). They estimated the effects of different profiles of adopters on the diffusion rate and ceiling for agricultural use of plastic sheeting, without, however, estimating the length of the innovation cycle. Several studies (Coughenour, 1961; Bishop and Coughenour, 1964; Deutschmann and

Hevens, 1965) investigated the reasons for discontinuing innovations, but no innovation cycle was estimated. Easingwood (1988) estimated product life-span patterns for new industrial products. His model provides an estimate for the overall life of a given technology from the first day of its appearance in the market until its final disappearance. Although the concept of technology cycle was implicitly included in Easingwood (1988), no use or estimate was provided.

Estimation of diffusion and abandonment processes for technologies used in the distant past may face problems of reliable data since documentation on the number (or share) of users may not be complete for the entire period. This may partially explain past difficulties in estimating technology cycles. The current study is fortunate to have data that allows detection of the diffusion-abandonment process. Several data points exist: (1) the number of users (or acres) and time (year) when the technology was first introduced, (2) present information on use of the technology, and (3) information on the number of new adopters during the diffusion phase only but not during the abandonment phase.

Diffusion and abandonment of a given technology in the absence of a complete data set for the stage of abandonment can be described using the concept of technology cycle that provides the rate at which technologies are being replaced. A technology cycle is defined here as the time period between the adoption of a particular technology by a decision maker and its abandonment or replacement by another technology. This concept has been used broadly in models of equipment replacement (e.g., Rifas, 1957, p. 67) that suggest replacement patterns for equipment used by identical producers. Therefore, the technology cycle hereafter presented is not a behavioral model (such as the model in Kislev and Shchori-Bachrach, 1973), but rather, a procedure to fit a curve to incomplete time series data. In doing so, one assumes that each technology is associated with a given lifespan (cycle) that does not change over time or as a result of market events. This is a simplifying assumption since technology cycles may be influenced by competing technologies, and prices, although, there are circumstances when this is not necessarily true (Dinar and Zilberman, 1991). Therefore, the estimates here can provide an upper bound to the technology cycle.

Let $t = t_0, \cdots, t_\tau, \cdots, t_T$ be the analyzed time period where t_0 and t_τ are the first and last years with observed number of adopters, and t_T is the last year in the sample for which the actual number of users of a particular technology is known (notice that $t_T \geq t_\tau$).

N_{t_T} is the observed number of users of the technology at time t_T, and n_t is the number of new adopters in year *t*. The variable $\boldsymbol{n_t}$ is defined as the cumulative number of adopters in year *t*, assuming (at this stage) no discontinuance or abandonment of that technology (pure accumulation).

Then:

(1) $$\boldsymbol{n}_t = \sum_{j=t_0}^{t} n_j$$

This definition of $\boldsymbol{n}_t$ accounts for the cumulative number of possible users as it appears in the data set. Since the above expression does not take into account the number of growers abandoning the technology, the value $\boldsymbol{n}_t$ at year t is not in agreement with the observed value for users of the technology in year t, and therefore:

(2) $$N_{t_T} \leq \boldsymbol{n}_{t_T}$$

and

(3) $$\boldsymbol{n}_{t_\tau} \leq \boldsymbol{n}_{t_T}$$

The variable N_t° is the estimated number of users of the technology at year t that is implicitly expressed as:

(4) $$N_t^\circ = f\left[n_{t_0},\ \sum_{j=t_0}^{t} n_j,\ z \right]$$

where z is the technology cycle (years).

More specifically, N_t° can be estimated as follows (see also Figure 13.1):

(5) $$N_t^\circ = \begin{cases} n_{t_0} & \\ N_{t-1}^\circ + n_t - \xi_t & \text{for} \\ N_{t-1}^\circ - \xi_t & \end{cases} \qquad \begin{matrix} t = t_0 \\ t_0 < t < t_T \\ t = t_T \end{matrix}$$

and

$$(6) \qquad \xi_t = \begin{cases} 0 & \text{for} \quad t - t_0 < z \\ 0 & \quad n_{t-z} = 0 \\ n_{t-z} & \quad t - t_0 \geq z \end{cases}$$

where ξ_t is the number of farmers abandoning the technology at year t. The estimated technology cycle (z^*) is then the value for the technology cycle that minimizes the difference between the observed and calculated number of users at year T:

$$(7) \qquad z^* = z \rightarrow \left\{ \min_z \left| N^{\circ}_{t_T} - N_{t_T} \right| \right\}$$

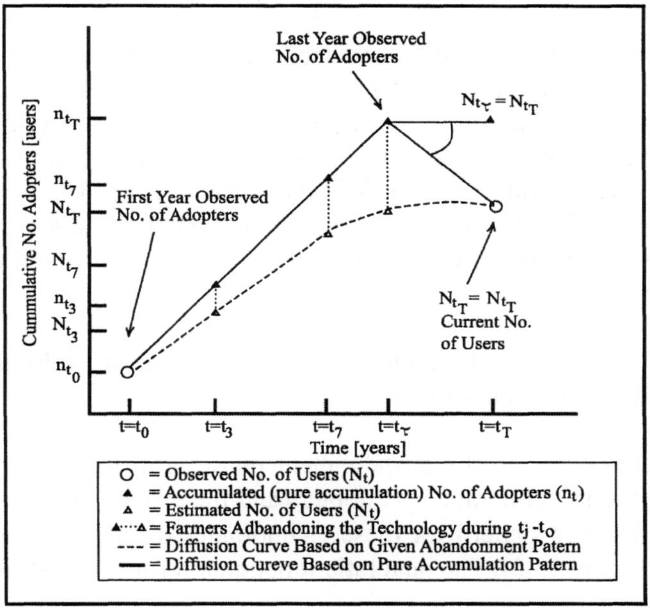

Figure 13.1. Scheme for the estimation procedure of the technology cycle

As a first step, technology cycles are estimated for each irrigation technology in every region using the system (4)-(7). This is done using a simulation program (simulated values for drag-line are presented in Figure 13.2) and a reasonable range of initial values for *z*. The chosen value (z^*) is the one that meets the criteria of equation (7).

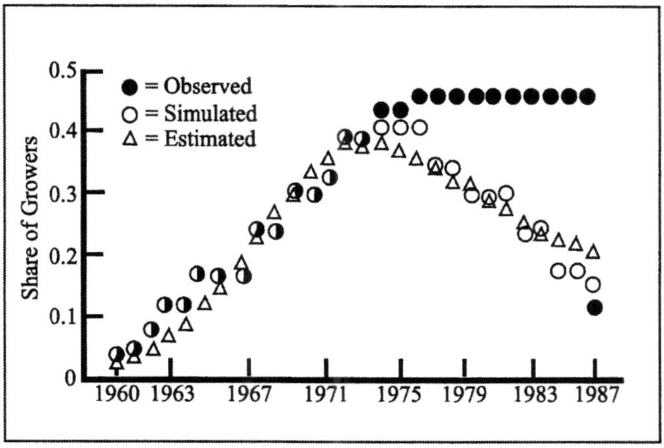

Figure 13.2. Observed, simulated and estimated values for the diffusion of drag-line sprinklers in Hadera region

To demonstrate the use of the technology cycle for creating the data needed to estimate a diffusion-abandonment curve for a given technology, assume for example, data for a 7-year period, and a cycle of 4 years. The number of new adopters each year is 5, 7, 4, 3, 4, 3, 0, over the period of the 7 years. The actual number of users will therefore be 5, 12, 16, 19, 18 (19 + 4 - 5), 14 (18 + 3 - 7), 11 (14 + 0 - 3). Notice that for the first 4 years, the technology was in a phase of diffusion and then for 3 years it is in an abandonment process.

2.2. Diffusion-Abandonment Curves for Irrigation Technologies that Have Been Abandoned

For technologies in the abandonment phase, a diffusion curve is developed using the initial number of adopters and the estimated technology cycle, z^*. The observed pattern for these technologies displays a quadratic curve over time (Figure 13.2). The quadratic functional form to be estimated for that process is:

(8) $$N_t + t/\left[b_1 t^2 + b_2 t + b_3\right]$$

In this expression N_t is the cumulative share of adopters by year *t*. The index t was normalized by setting the first year of the diffusion at 1. b_1, b_2, and b_3 are the estimated coefficients. First-order conditions with respect to t provide the estimate for the year (t^*) in which the ceiling was reached:

(9) $t^* = (b_3/b_2)^{1/2}$

Substituting t^* into the quadratic diffusion curve (8) yields the estimated ceiling, N_{t^*}. For $t > t^*$ the diffusion process is negative, meaning that the technology is being abandoned. For technologies that are still in a process of abandonment, the year of complete abandonment can be estimated by solving equation (8) with the estimated values for the b_i's, and setting N_t to be zero.

2.3. Effect of Input and Output Prices on the Diffusion of Modern Irrigation Technologies

A number of studies have recognized that diffusion of a technology could be affected by the product price and the profitability expected to result from the technology. Griliches (1960) showed that the rate at which growers accept a new agricultural technology depends, among other things, on the magnitude of the profit to be realized from the changeover. Mansfield (1963) showed that diffusion rates of new technologies in several industries were positively related to the profitability of those technologies. Jarvis (1981) showed that both the rate and the extent of diffusion for new technologies were positively related to the profitability of those technologies. Using a normative model, Dinar and Letey (1989) demonstrated the positive combined effects of charges for irrigation water and capital subsidies for irrigation technologies on the economics of technology selection under various limiting environmental conditions. Theoretical and empirical evidence cited in Feder et al., (1985) and in Thirtle and Ruttan (1987), provides the basis for developing an empirical model to estimate effects of output prices and input costs on the diffusion of technologies. Long-term investment decisions, such as in irrigation equipment, are based on past and future price expectations for inputs and outputs. For the purpose of our analysis assume that only past prices of input and output (in year t - t) affect the decision to invest in irrigation equipment.

This section demonstrates the use of economic variables to estimate diffusion curves. These variables represent crop yield price, water price, and subsidies provided by the government for the investment in irrigation technologies. (The model is applied to technologies that are still in the diffusion phase.)

The basic logistic equation for the diffusion process is:

(10) $N_t = d_1/[1 + exp(-d_2 - d_3 t)]$

where d_3, which is the rate of diffusion, is assumed to be a function of crop profitability, water price, and the subsidy for the investment in the technology (capital cost). By expressing d_3 as a linear function of these variables ($d_3 = \psi_0 + \psi_3 P_{t-t} + \psi_4 P^W_{t-t} \psi_5 S_{t-t}$), the basic logistic equation becomes (Jarvis, 1981):

$$(11) \qquad N_t = d_1' / \left[1 + \exp\left(-d_2' - \psi_0 t - \psi_3 t P_{t-\mathrm{t}} - \psi_4 t P^W_{t-\mathrm{t}} - \psi_5 t S_{t-\mathrm{t}}\right)\right]$$

where $P_{t-\mathrm{t}}$ is a variable measuring crop profitability, $P^W_{t-\mathrm{t}}$ is water price, and $S_{t-\mathrm{t}}$ is a variable measuring subsidy level for the capital cost of the technology in year ***t* - *t***.

3. DATA AND EMPIRICAL SPECIFICATIONS

The models presented earlier were applied to data from a study of citrus groves in Israel. The study sample includes only groves, which are owner-operated, and are greater than 2.5 ha (1 ha = 2.5 acres or 10 dunams). Excluded are groves operated either by cultivation companies on the basis of fixed payment, or by part-time operators[1]. These kinds of operators were observed to be motivated by economic considerations extremely different than full time owner operators (e.g., Guttman and Haruvi, 1986; Feder et al., 1988). A total of 209 groves owned by *kibbutz* (collective settlement), *moshav* (cooperative settlement), and private owners were sampled. These groves are from settlements in six regions (from north to south: Hadera, Ra'anana, Rehovot, Lackish, Negev and Gaza Strip). The sampled area accounts for 16 percent of the total citrus area in Israel (Table 13.1). Questionnaires were completed during the course of interviews conducted with growers between October 1986 and April 1987. General information on sample size and current distribution of irrigation technologies by regions is presented in Table 13.1.

Irrigation technologies in common use during the study period were (in order of their introduction to the market): (1) traditional irrigation such as border and furrow, (2) hand-moved sprinklers (aluminum pipes), (3) solid set sprinkers above canopy (hereafter referred to as 'above-canopy', (4) drag-line sprinklers under canopy (plastic pipes), (5) solid-set sprinklers under canopy (plastic pipes), (6) low volume micro-sprinklers and micro-jets, and (7) drip irrigation. Farmers in most of the regions have abandoned the first four irrigation technologies. The later three technologies are still in a process of diffusion in most of the regions. These three technologies will be identified hereafter as 'the modern technologies'. Additional information with regard to the data, as well as detailed technology characteristics and associated costs, can be found in Dinar and Yaron (1988). Information regarding number of groves and adoption periods for various technologies by region are presented in Table 13.1.

Table 13.1. Characteristics of sample citrus farms, 1987, by region

	Region					
	Gaza	Negev	Lackish	Rehovot	Ra'anana	Hadera
Citrus Area (1000 ha)	[a]	2.9	3.6	11.4	11.0	6.5
Sample area (ha)	375	1873	633	1030	1210	759
Sampled groves	44	57	21	28	25	34
Technology	Percent of area equipped with technology					
Traditional (furrow)	50.3	0	0	0	0	0
Hand-moved	0	0	0	0	0	0.7
Above-canopy	0	0	0	0	2.3	7.9
Drag-line	0	32.0	40.5	21.4	23.4	8.3
Solid-set	5.6	5.4	28.9	27.8	35.9	42.5
Micro-sprinkler	44.1	7.0	17.7	48.6	34.3	38.2
Drip	0	45.6	12.0	2.3	4.1	2.4

[a]Aggregated data not available

The period of interest for purposes of this study began in the fifties (although data for some technologies exist from the beginning of the century), when hand-moved sprinker replaced the traditional furrow irrigation in some established groves and also became the irrigation technology that was introduced in newly established groves. The hand-moved sprinkler system, consisting of aluminum pipes, was labor intensive and allowed for very little flexibility with regard to irrigation schedule.

Table 13.2. Plantation and adoption periods for irrigation technologies in different regions in the sample

Region	Gaza	Negev	Lackish	Rehovot	Ra'anana	Hadera
Plantation period	1930-68	1954-74	1954-64	1932-79	1920-78	1901-67
	Observed adoption period					
	$t_0 - t_\tau$	$t_0 - t_\tau$	$t_0 - t_\tau$	$t_0 - t_\tau$	$t_0 - t_\tau$	$t_0 - t_\tau$
Furrow	1930-68			1932-57	1920-54	1901-61
Hand-moved		1954-64	1954-64	1950-63	1946-64	1946-67
Above-canopy				1965-68	1960-72	1949-77
Drag-line		1960-73	1962-72	1960-76	1962-75	1960-77
Solid-set	1977-78	1961-81	1970-80	1965-80	1970-76	1960-84
Micro-sprinklers	1975-83	1970-85	1975-81	1973-85	1970-83	1970-84
Drip		1967-86	1970-81	1972-82	1978-81	1976-85

t_0 = first year with observations on adoption

t_τ = last year with observations on adoption

Drag-line systems which were first introduced in the early sixties have been in use for the longest period of time, although this technology is labor-intensive and very difficult to control. The introduction of solid set sprinklers both above and under the canopy saved labor and contributed to better control of water application to individual trees. Both of these technologies however, are capital intensive and labor extensive in comparison to the traditional furrow irrigation. Other disadvantages associated with the solid set technologies involve operational difficulties, irrigation water uniformity, and salinity problems (in the case of the above-canopy).

Micro-jet and micro-sprinkler systems (hereafter referred to as micro-sprinklers) introduced in the seventies are capital intensive, but require only low water pressure, save labor, provide better irrigation water uniformity, and are easy to control. Drip irrigation systems demonstrate the same advantages as micro-sprinklers, and are also less capital intensive than micro-sprinklers.

For the purpose of estimating the effect of economic variables on diffusion, three variables were used. The first variable is the export price ($P_{t\text{-}t}$) for the *shamuti* variety of citrus (US$ per 10 kg), on the assumption that *shamuti* prices represent other citrus variety prices (*shamuti* is also the main crop in the database). The information for constructing this variable was collected from data recorded in the Statistical Abstract of Israel (various years) with values represented in 1984 constant US dollars. The second explanatory variable is water price ($P^W_{t-\mathrm{t}}$), calculated in US$/m^3 (1233.5 m^3 = 1 acre-foot) from the Statistical Abstract of Israel (various years) assuming the same price for all regions. The assumption is quite reasonable under conditions prevailing in Israel since a central authority, which does not discriminate among regions, dictates water prices. The third variable is the subsidy rate ($S_{t\text{-}t}$) on government loans for irrigation equipment (Israel Ministry of Agriculture, various years). These rates may differ by technologies, but it is assumed that no difference exists between regions. This assumption holds for the regions included in the database, but not necessarily for other regions, which may receive preferred subsidy rates.

For the purpose of estimating the diffusion logistic curves for solid set, micro-sprinklers, and drip, no distinction is made among regions and all 209 observations are grouped in one data set.

4. RESULTS

Data on the current shares of the different irrigation technologies in various regions are presented in Table 13.1. In 1987, more than 50 percent of the grove area (groves and growers may be used hereafter in the same context) was equipped with modern irrigation technologies. The diffusion processes of these technologies began in the early sixties to the early seventies, depending on the region and the irrigation technology (Table 13.2). Hand-moved sprinklers were adopted in the fifties in five regions, but by 1987 this system was no longer in use (except for less than 1 percent of the area in Hadera). Above-canopy sprinklers were adopted in the three northern regions (Rehovot, Ra'anana, and Hadera), but are found today in only 2.3 and 7.9 percent of the grove area in the Ra'anana and Hadera regions, respectively. Drag-line sprinklers were adopted by growers in five regions but are used today on only 32 percent of the citrus groves in the Negev, 40.5 percent in Lackish, 21.4 percent in Rehovot, 23.8 percent in Ra'anana, and 8.3 percent in Hadera.

For each irrigation technology with declining use over time, data are available on the number of groves currently farmed by that technology (as of 1987), and annual number of adopters from as early as 1901 to 1987. The missing piece of information relates to the number of growers abandoning the technology each year. The procedure developed to estimate the technology cycle was applied to the data on furrow, hand-moved, above-canopy, drag-line, solid-set, and micro-sprinklers for each region separately. (The technology cycle cannot be applied to technologies still in a diffusion process using the approach developed here.) Results for the technology cycle estimates are presented in Table 13.3. The estimated technology cycle for furrow irrigation is 26-30 years; the cycle for hand-moved sprinklers is 22-24 years, for above-canopy sprinklers it is 26-28 years; for drag-line sprinklers it is 17-20 years; for solid-set sprinklers it is 17 years (only in the Negev); and for micro-sprinklers the cycle is 15-17 years (only for the Negev and Hadera Regions). In general, the Negev and Lackish regions exhibit shorter technology cycles for all irrigation technologies than the other regions. However, these differences were not found to be statistically significant.

Table13.3. Estimated irrigation technology cycles (years)

Tech.	Gaza	Negev	Lackish	Rehovot	Ra'anana	Hadera
Furrow	28	a	a	30	26	26
Hand	a	22	23	24	22	23
Above	a	a	a	26	27	28
Drag	a	18	18	20	19	17
Solid	a	17	b	b	b	b
Micro	*	15	b	b	b	17
Drip	a	b	b	b	b	b

[a]Not in use
[b]In the diffusion process

The number of farmers using a particular irrigation technology being renounced was calculated for each region using the technology cycle. Then quadratic logistic equations were estimated using a non-linear, iterative, least squares procedure (SAS, 1985). Since there is a tremendous volume of information, only results for the Rehovot and Hadera regions are presented in Table 13.4. A curve depicting the estimated diffusion and abandonment of drag-line sprinklers in the Hadera region is presented in Figure 13.2. Results for all regions and technologies can be found in Dinar and Yaron (1988).

The coefficients presented in Table 13.4 (and additional coefficients that are not presented) were used to estimate the year when the diffusion process reached its ceiling for different technologies. Application of the procedure (equation 9) to the drag-line sprinkler equation (Table 13.4), indicates that the ceiling was reached 15-19 years after the beginning of the diffusion process (depending on the region). For technologies in the process of being renounced (but not yet abandoned), the coefficients in Table 13.4 also make it possible to estimate, the year when a technology will be fully abandoned.

Table 13.4. Estimated logistic quadratic diffusion and abandonment curves for technologies being abandoned, by regions[a]

	Irrigation technology			
	Furrow	Hand-moved	Above-canopy	Drag-line
		Rehovot region		
R^2	0.865	0.800	0.565	0.957
b_1	0.403	0.168	0.995	0.132
	(0.054)	(0.025)	(0.284)	(0.016)
b_2	-25.15	-4.14	-2.77	-3.59
	(3.817)	(0.857)	(5.268)	(0.582)
b_3	482.09	40.92	61.29	46.64
	(66.263)	(7.057)	(21.998)	(4.926)
		Hadera region		
R^2	0.908	0.836	0.764	0.926
b_1	0.430	0.237	0.819	0.444
	(0.031)	(0.032)	(0.262)	(0.049)
b_2	-36.91	-7.24	-25.10	-11.02
	(2.997)	(1.259)	(9.693)	(1.556)
b_3	916.86	76.91	333.62	103.91
	(70.943)	(11.924)	(86.623)	(11.923)

In parentheses are asymptotic standard deviations of the coefficients
[a]Results are presented for only two regions

For the drag-line sprinklers that were still in use on small portions of groves in the various regions, it is estimated (not presented) that this technology will disappear 30-35 years after initial adoption. Specifically, it is estimated that during the year 1990, drag-line systems will no longer be used in the Negev, Lackish, and Hadera regions; and in 1995 they will also disappear from the Rehovot and Ra'anana regions.

The three 'modern technologies'—solid-set sprinklers, micro-sprinklers and drip systems—are used in all regions (except for Gaza where drip was not used). Therefore, aggregated logistic diffusion curves were estimated for the sample data. A range of lag periods from 1 year to 5 years was used in the analysis (not presented), however, a lag of one year provided the most reasonable results. A one-year lag is therefore used here, assuming that a decision regarding the installation of a technology in year t depends upon conditions existing in year $t-1$. Some difficulties were encountered in estimating these logistic curves. The Durbin-Watson (D.W.) statistic is, however, in the intermediate range, indicating inconclusive results with regard to positive serial correlation. Draper and Smith (1981) suggest such cases be treated as if a serial correlation had been found. Because of these difficulties, the coefficients are not presented. In order to correct for the possible presence of first order serial correlation, the Hildreth-Lu procedure (Pyndick and Rubinfeld, 1981) was applied. The corrected logistic expression is now:

$$(12) \quad N_t = d_1'/\left[1 + exp\left(-d_2' - \psi_0 t - \psi_3 t P_{t-1} - \psi_4 t P_{t-1}^W - \psi_5 t S_{t-1}\right)\right]$$

where

$\boldsymbol{N}_t \quad = N_t - \gamma N_{t-1}$
$\boldsymbol{P}_{t-1} \quad = P_{t-1} - \gamma P_{t-2}$
$P_{t-1}^W \quad = P^W{}_{t-1} - \gamma\, P_{t-2}^W$

and

$\boldsymbol{S}_{t-1} \quad = S_{t-1} - \gamma S_{t-2}$

and γ is a scalar of grid values($0 \le \gamma \le 1$).

Table 13.5. Logistical curves of the diffusion of several modern irrigation technologies (corrected for serial correlation)

Dependent variable:	Share of groves using technology		
	Irrigation technology		
	Solid-set	Micro - sprinkler	Drip
γ	0.2	0.3	0.4
Asymptotic R^2	0.997	0.997	0.995
d_1'	0.358	0.561	0.234
	(0.003)	(0.006)	(0.003)
d_2'	-4.93	-8.26	-6.34
	(0.178)	(.279)	(0.246)
ψ_0	0.352	0.419	0.308
	(0.014)	(0.014)	(0.013)
ψ_3 (yield price)	0.00044	0.00300	0.00310
	(0.0012)	(0.0008)	(0.0010)
ψ_4 (water price)	0.00052	0.00481	0.00550
	(0.0018)	(0.0013)	(0.0021)
ψ_5 (subsidy)	0.01	0.03	0.05
	(0.0063)	(0.0072)	(0.0099)
D.W.	2.01	1.95	1.98
Actual share of adopters at 1987	0.35	0.51	0.22

In parenthesis are symptomatic standard deviations

Equation (12) was estimated for a range of γ values using the same procedure as equation (10). The chosen value for γ is that given by the smallest sum of squared residual for the regression runs. The chosen values of γ are presented in Table 13.5 for each technology; an autoregressive transformation was performed and the results of the regression runs are also presented. Values for the D.W. statistic are now higher than in the original regression, indicating that no serial correlation exists in the transformed estimated residual.

In all cases the estimated ceiling (d_1') is higher than the actual share of adopters in 1987, indicating that the diffusion process will reach its ceiling after that year. All coefficients affecting the diffusion process behave as reported in the literature: an increase in *shamuti* export price, in water price, and in subsidy rate for modern irrigation equipment, will increase the share of the modern technologies used in citrus groves. These findings are in agreement with results provided by Caswell and Zilberman (1985). The speed at which the ceiling will be reached can also be influenced by those variables. For example (not presented in the tables), if export prices remain at 1987 level while holding all other variables constant, the ceiling will be reached in the years 1999, 2000, and 2001 for solid-set, micro-sprinklers, and drip, respectively. If export prices increase by 10 percent the ceiling will be reached one year earlier than in the previous case for all technologies. A decrease of the same rate in export price will result in a one-year delay in approaching the ceiling for all technologies.

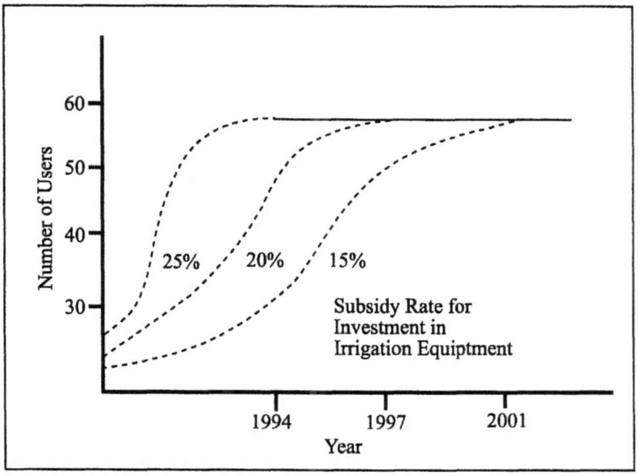

Figure 13.3. Effect of subsidy for irrigation equipment on the diffusion of drip (national data)

There are some tradeoffs between greater use of modern technologies, or shorter adoption periods, and social cost. For example, in the case of externalities in the production process related to irrigation, a regional authority might be interested in transition to more efficient irrigation technologies. This change is associated with additional investment that might be subsidized by society through tax dollars, and should therefore be evaluated in this regard. The effect of changes in subsidy rates for irrigation equipment on diffusion of drip is demonstrated in Figure 13.3. A subsidy rate for irrigation equipment of 20 percent would result in reaching the ceiling four years earlier (1997 instead of 2001) than with a subsidy rate of 15 percent, where as with a subsidy rate of 25 percent, the diffusion reaches the ceiling in 1994. The time gained by the increased subsidy can be weighted against possible losses resulting from continued use of the existing irrigation technology.

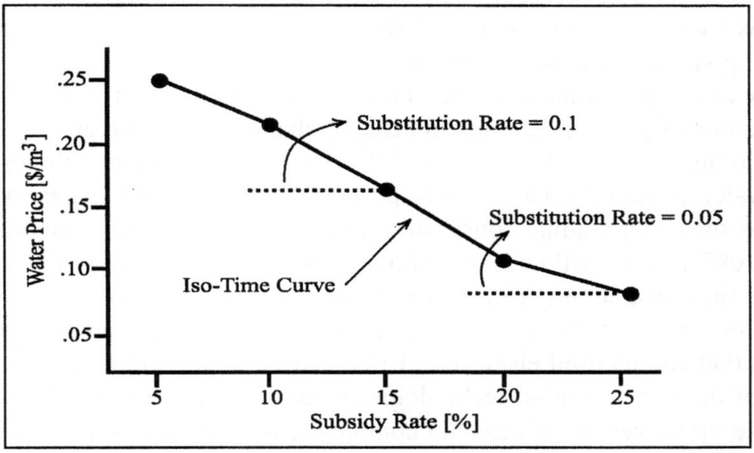

Figure 13.4. Substitution between water price and subsidy for irrigation equipment in order to reach the ceiling of the diffusion process for drip at year 2000 (national data)

Another demonstration of the usefulness of the results is presented in Figure 13.4. Here two policy variables are substituted in order to achieve the ceiling of the diffusion process for drip irrigation at year 2000: water price and subsidy rate. Both can be controlled by the government and used for policy purposes. By drawing a substitution curve between these two policy variables (using the estimated equation for drip irrigation in Table 13.5), it was found that the substitution rate is larger in cases where water prices are high (substitution rate of 0.1) than in cases of low water prices (substitution rate of 0.05). These findings can, therefore, serve the policymaker in optimizing the combination between these two variables.

5. CONCLUSIONS

In this study adoption and abandonment processes of irrigation technologies in citrus groves were estimated and depicted. Data from Israel and Gaza were used to: (1) estimate technology cycles for different irrigation technologies,

(2) estimate, using the technology cycle, the process of diffusion-abandonment of technologies already in the process of renouncement, and (3) confirm hypotheses established in previous theoretical and empirical studies with regard to the effects of input and output prices on the diffusion processes of irrigation technologies still in the diffusion phase.

It was found that the technology cycle period for a given technology is dependent only on the technology and not on physical conditions prevailing in the different regions (e.g., weather, soil types etc.). For technologies being abandoned, the technology cycle was used to estimate the year of discontinuance. These findings can serve policymakers of developing agricultural regions, as well as manufacturers of irrigation equipment who are interested in predicting years of use for a given technology.

In many cases, policymakers must consider the effects of possible policies on the behavior of growers in order to achieve changes in resources use. For example, in the United States, the new 1990 Farm Bill considers policies to improve water conservation and reduce pollution problems. In Asia, as irrigation water becomes the binding constraint for rice production, improvements in irrigation efficiencies will need to be found. Studies suggest that farmers should improve their irrigation performances by transition to modern irrigation practices. The current study, although based on data from one country, provides insights that can be used elsewhere. Effects of input and output prices on diffusion of modern irrigation technologies were estimated and used to demonstrate the effectiveness of possible combinations of policy variable levels on achieving a range of technology diffusion rates. These variables were found to be very effective in determining the rate and ceiling of the diffusion process.

ACKNOWLEDGEMENT

This study was supported in part by the U .S. Agency for International Development to the Hebrew University of Jerusalem as a part of a Tri-national Egypt-Israel-USA Project “Patterns of Agricultural Technology Exchange and Cooperation in a Similar Ecosystem: The Case of Egypt and Israel.”

The specialists of the Agricultural Extension Service in Israel and Gaza provided help in data collection.

REFERENCES

Bishop, R. and C. M. Coughenour. Discontinuance of farms innovations. *Bulletin of Agricultural Economics, 361.* Department of Agricultural Economics and Rural Sociology, Ohio State University. Columbus, OH. 1964

Cameron, H. M. and J. S. Metcalfe. On the economics of technological substitution. *Technology, Forecasting and Social Change,* 32:147-162. 1987.

Casterline, G., A. Dinar and D. Zilberman. The adoption of modern irrigation technologies in the United States in: "Free Trade and Agricultural Diversification in Canada and the United States" (A. Schmitz, ed.). Westview. Boulder, CO. 222-248. 1989.

Caswell, M. and D. Zilberman. The choices of irrigation technologies in California. *Americam Journal of Agricultural Economics*, 67:224-234. 1985.

Coughenour, C. M. The practice-use tree and the adoption, drop-out, and non-adoption of recommended farm practices: a progress report. Paper presented at Rural Sociological Society Meeting. Ames, lA. 1961.

Deutschmann, P. J. and A. E. Hevens. Discontinuances: a relatively uninvestigated aspect of diffusion. Department of Rural Sociology, University of Wisconsin. Madison, WI. 1965.

Dinar, A. and J. Letey. Economic analysis of charges and subsidies to reduce agricultural drainage pollution. Proceedings if the 2nd pan-American regional conference. "Toxic Substances in Agricultural Water Supply and Drainage - An International Environmental Perspective" (J. B. Summers, ed.). Ottawa, Ontario, Canada. U.S. Committee on Irrigation and Drainage. Denver, CO. 53-70. June 8-9, 1989.

Dinar, A. and D. Yaron. Adoption of new irrigation technologies in citrus: process description and the influence of inputs scarcity and quality. Tri-national Egypt-Israel-USA Project, Patterns of Agricultural Technology Exchange and Cooperation in "Similar Ecosystem: The Case of Egypt and Israel." 1988.

Dinar, A. and D. Zilberman. Effects of input quality and environmental conditions on selection of irrigation technologies in: "The Economics and Management of Water and Drainage in Agriculture" (A. Dinar and D. Zilberman, eds). Kluwer Academic Publishers. Boston, MA. 229-250. 1991.

Draper, N. R. and H. Smith. "Applied Regression Analysis" (2nd Edition). Wiley, New York. 1981.

Easingwood, C. J. Product life cycle patterns for new industrial products. *R & D Management*, 18:23-32. 1988.

Feder, G., R. E. Just and D. Zilberman. Adoption of agricultural innovations in developing countries, a survey. *Economic Development and Cultural Change,* 33:255-298. 1985.

Feder, G., T. Onchan, Y. Chalamwong and C. Holgladarom. "Land Policies and Farm Productivity in Thailand." Johns Hopkins University Press for the World Bank. Baltimore, MD. 1988.

Fishelson, G. and D. Rymon. Adoption of agricultural innovation: the case of drip irrigation of cotton in Israel. *Technological Forecasting and Social Change*, 35:375-382. 1989.

Fisher, J. C. and R. H. Pry. A simple substitution model for technological change. *Technological Forecasting and Social Change,* 2:75-84. 1971.

Grilliches, Z. Hybrid corn and the economics of innovations. *Science*, 132:275-280. 1960.

Guttman, J. M. and N. Haruvi. Cooperating and part-time farming in the Israeli moshav. *American Journal of Agricultural Economics*, 68:77-87. 1960.

Hayami, Y. and V. W. Ruttan. "Agricultural Development: An International Perspective." Johns Hopkins University Press. Baltimore, MD. 1971.

Israel Ministry of Agriculture. Bureau of the Water Department for irrigation efficiency. Price Lists and Loan Terms. Tel Aviv, Israel. Various years. (Hebrew)

Jansen, H. G. P., T. S. Walker and R. Barker. Adoption ceilings and modern coarse cereal cultivars in India. *American Journal of Agricultural Economics*, 72: 653-663. 1990.

Jarvis, L. S. Predicting the diffusion of improved pasture in Uruguay. *American Journal of Agricultural Economics*, 63:496-502. 1981.

Kislev, Y. and N. Shchori-Bachrach. The process of an innovation cycle. *American Journal of Agricultural Economics,* 55:28-37. 1973.

Kulshreshtha, S. N. Irrigation and prairie agricultural development in: "Free Trade and Agricultural Diversification in Canada and United States" (A. Schmitz, ed.). Westview. Boulder, CO. 222-248. 1989.

Mansfield, E., Intrafirm rates of diffusion of an innovation. *Review of Economics and Statistics*, 45. 1963.

Pindyk, R. S. and D. L. Rubinfeld. "Econometric Models and Economic Forecasts" (2nd Edition). McGraw-Hill, New York. 1981.

Rifas, B. E. Replacement models in: "Introduction to Operations Research" (C. W. Churchman, R. L. Ackroff and E. L. Arnoff, eds.). Wiley, New York. 1957.

Rogers, E. M. Diffusion of Innovations (3rd Edition). Free Press. New York. 1985.

SAS. SAS User's Guide Version, 5 Edition. SAS Institute Inc. Cary, NC. 1983.

Statistical Abstract of Israel. Central Bureau of Statistics. Jerusalem, Israel. Various years.

Thirtle, G. C. and V. W. Ruttan. The role of demand and supply in the generation and diffusion of technical change in: "A Volume of the Economics and Technological Change Section" (F. M. Scherer, ed). Harwood Academic Publications. New York. 1987.

[1] It should be noted at the estimated share of these groves was 40 percent of the total area, but their share in production was less than 15 percent due to bad maintenance. Also, many of these groves went out of production after 1987 (N. Ravid, Head, Extension Service Department of Citrus, Hakiria Tel Aviv, personal communication, 1990).

14

AN APPROACH TO THE PROBLEM OF WATER ALLOCATION TO ISRAEL AND THE PALESTINIAN ENTITY*

Dan Yaron
The Hebrew University of Jerusalem, Rehovot, Israel

1. INTRODUCTION

The paper is composed of two parts. The first part presents the water scarcity problem in the region. It reviews the water supply-demand situation, with reference to Israel and the Palestinian Entity. Means of water saving are discussed and supply augmenting options are briefly mentioned. In the second part, the problems of interentity allocation of the limited water resources are discussed and an approach towards finding a solution is suggested.

2. WATER SCARCITY AND ITS DIMENSIONS

The countries of the region face severe scarcity in their freshwater supply potential, which involves both quantity and quality dimensions.

2.1. Israel

In Israel nearly all the freshwater supply potential is being utilized. With the urban demand continuously increasing (domestic consumption) a n ever increasing share of treated wastewater is allocated to agriculture—30 and 50 percent are projected for the years 2000 and 2010, respectively. Also, an increased quantity of brackish water is planned for agricultural use. The water supply-demand balance in Israel is presented in Table 14.1. In this table two

*Permission to print this chapter was granted by Elsevier Science. The chapter was originally published in *Resources and Energy Economics*, 16:271–286, 1994.

estimates of natural sources potential are presented. The higher one is cited from Tahal Master Plan (Tahal, 1988), which includes potential sources, the utilization of which requires considerable investment. The lower estimate is based on Nvo (1992) who provided a more realistic interpretation of the potential. The demand figures are taken from Tahal Master Plan (Tahal, 1988), with projected population of 5.7 and 6.7 million in the years 2000 and 2010, respectively. Note that updated population projections for 2010 are higher in view of the large wave of immigration to Israel in 1990 and 1991.

Table 14.1. Projected water supply potential from natural sources (fresh and brackish) and projected use in Israel in the years 2000 and 2010[a]

Source	Low [b]	High [c]
A. Supply MCM/year		
Groundwater	950	1115
Lake Kinereth Basin	600	660
Floodwater	30	80
Total	1580	1855
Conveyance losses	63	74
Net supply potential	1517	1781
	Year	
	2000	2010
B. Demand MCM/year		
Domestic [d]	642	801
Industrial	134	148
Agricultural	1184	1070
Total	1960	2019
C. Difference		
B-A	189-443	238-602

Source: Tahal, (1988) and Nvo (1992)
[a]Not including treated wastewater
[b, c]The higher estimate is Tahal's (1988), which includes theoretical potential, the utilization of which depends on considerable investment. The lower estimate is by Nvo (1992), which appears to be more realistic in view of the developments since 1988.
[d]Including settlers in the West Bank and Gaza Strip

The gap between the projected demand and supply from natural sources (B-A) is to be filled by treated wastewater and some local desalination of brackish water.

Of major concern is the quality issue. For example, in the Coastal Aquifer, a major source of tap water supply, the salinity is rising at the rate of 2.5-3 ppm chlorides per year. Continuous salinization of numerous water wells is observed in other regions.

Increasing use of treated wastewater in irrigation, and a high level use of chemicals in agriculture brings the focus to environmental problems and the protection of natural resources: land and water.

Strict restrictions on the use of treated wastewater in irrigation have been declared; the issue of gradual contamination of groundwater by chemicals is being discussed; however, problems of enforcement of the regulations are far from being solved.

Another major problem is the reliability of the water supply over the years. It arises due to the annual fluctuations in rainfall and groundwater replenishment on the one hand, and low long-run water storage capacity, on the other.

The overall situation in Israel is one of a hardly balanced supply and demand situation with severe water quality problems. A widely accepted operational conclusion is that the use of treated wastewater must be planned and managed by one institution. The realization of this conclusion still faces organizational and bureaucratic difficulties.

The bottom line of this overview is that water is not a homogeneous resource. It is rather a multi-dimensional substance, which implies a clear-cut distinction among (a) high quality (fresh) water; (b) treated wastewater; (c) brackish water; (d) location of the source of supply; and (e) the timing of the supply.

2.2. West Bank and Gaza Strip

The natural source of water supply for the West Bank is the Mountain Aquifer, which can be schematically sub-divided into three subaquifers with estimated safe yields as given in Table 14.2.

The Mountain Aquifer extends across the pre-1967 "Green Line" border between Israel and the West Bank. Historically most of its potential was used by Israel; a considerably smaller share was used by the West Bank, mainly from the Eastern and Northern subaquifers.

The safe yield of the coastal subaquifer, which supplies water to the Gaza Strip is estimated at 60 MCM per year. The current use is estimated at 120-150 MCM/year (Awartani, 1990; Kally, 1989; Soffer, 1992), leading to an annual overuse deficit of 60-90 MCM/year. Due to the excessive overuse and the resulting intrusion of seawater, the acquifer's groundwater is gradually deteriorating and its salinity content is rising (currently 2/3 of the 60 MCM potential contains 400 ppm Cl; Goldberger, 1992).

Projections of water requirements of the West Bank and the Gaza Strip according to Awartani (1990) for the year 2000 are presented in Table 14.3.

An extrapolation of Awartani's projections leads to an estimated demand in the West Bank and Gaza Strip in the range of 450-500 MCM/year in 2010 due to the expected increase both in population and per capita domestic consumption.

Table 14.2. Water supply potential of the Mountain Aquifer

	Safe yield (MCM/year)	Of this, brackish water (MCM/year)
Western subaquifer (Yarkon-Tanimin)	350	40
Eastern subaquifer	200	100
North eastern subaquifer	130	100
Total	680	240

Table 14.3. Population and water requirements projections for the West Bank and the Gaza Strip for the year 2000 according to Awartani (1990)

Region	Population (thousands)	Water use MCM/year			
		Domestic	Industrial	Agriculture	Total
West Bank	1,776[a]	101	29	113	243
Gaza Strip	885	52	10	84	146
Total	2,661	153[b]	39[c]	197	389

[a]Assuming 300,000 Palestinians returning to the West Bank after a political settlement is reached.
[b]Not including visitors
[c]The total of 39 MCM is projected by Awartani. The subdivision between the West Bank and Gaza Strip is based on the author's estimate.

Considerably lower projections of West Bank and Gaza Strip water requirements for the years 2000 and 2010 were formulated by Schwartz (Tahal, 1990). The major source of the discrepancy between Schwartz and Awartani (1990) are population projections and the extent of agricultural demand. Both assume gradually rising levels of domestic per capita consumption with the 2010 level being still well below the levels projected for Israel.

2.3. Jordan

In Jordan, severe water scarcity prevails (for details see Soffer, 1992), and Jordan cannot be considered a source of relief for the Israeli-Palestinian water shortage. On the contrary, it is a source of potential demand for a share in any agreement on water allocation and use in the region at large.

2.4. Regional Water Quantity, Quality, and Scarcity

In summary, the joint potential supply of water for Israel, the West Bank and Gaza Strip is quite short of the projected needs. The sources of the potential conflict shortage are both in terms of quantity, quality and resource conservation. The West Bank is located above the Mountain Aquifer, which is separated from the mountainous surface of land by calcareous rocks with cracks through which biological and chemical contaminants may penetrate into the aquifer. Accordingly, most of the water allocated to the West Bank should be of high quality, in order to preserve the water quality of this aquifer in the long run. Furthermore, cooperative Israel and the Palestinian entity conservation oriented solutions for sewage disposal are necessary. Since such solutions are quite costly, this is a potential source of friction between the future Palestinian entity and Israel.

3. COUNTERMEASURES TO SCARCITY

Water saving technologies in agriculture can lead to a reduction in the demand for water through substitution of capital for water. Modern pressurized irrigation technologies can save 30-35 percent of water per land unit area, in comparison with the traditional surface irrigation and especially so in comparison with "zig-zag" furrows. An additional 25-30 percent can be saved in the conveying system. A study of the modernization of a traditional irrigation water project in the Jiftlik area in the West Bank indicated a considerable saving of water per ha in the modernization project, which at the same time, was accompanied by a rise in yields and income (Yaron and Regev, 1989; Regev et al., 1990). Some examples are given in Table 14.4.

Table 14.4. Water inputs and yields in the traditional and the modernization project

Crop	Traditional		Modernized		Effects of modernization	
	Water input (m^3/ha)	Yield (ton/ha)	Water input (m^3/ha)	Yield (ton/ha)	Water saving (%)	Yield increases (%)
	(1)	(2)	(3)	(4)	(5)	(6)
Early beans	7000	7.5	4000	10	43	33
Squash	8000	10.0	5000	20	38	100
Tomato	11000	15.0	7550	50	31	233
Hot pepper	11000	6.0	8000	8	27	333
Eggplant	12000	15.0	8450	60	30	300

Source: Agricultural Planning and Development Office, West Bank, and Irrigation Extension Specialists, Israel Ministry of Agriculture, cited from Regev et al. (1990).
Notes: (5) ={ [(1) – (3)]/(1)} × 100; (6) = {[(4)-(2)}/(2)} × 100

It should be noted that: 1) The modernization of the Jiftlik project included the frontier elements of new technologies such as fertigation, drip irrigation and installation of on-farm operational reservoirs, which made it possible to supply water to the fields whenever needed by the plants instead of the inflexible strict biweekly supply from the water tributaries according to the traditional water rights; 2) The modernization of the Jiftlik irrigation project was accompanied by a shift to a new mix of crops (mainly high income vegetables for export) and to a modern production technology package involving high yield varieties, plastic covering of land for the early part of the growing period and the use of pesticides; 3) Under the traditional irrigation project more than 30 percent of the gross land area was consumed by dirt canals and "zig-zag" furrows; the modernized project increased considerably the net cultivated area.

The results of the analysis of the Jiflik project showed also that the link between the modernization of the irrigation system and a new mix of crops is necessary for economic feasibility; modernization of the irrigation system *alone* without the new mix of crops would not have been economically justified. (The traditional mix comprised of 213 kinds of grains and 113 vegetables, all traditional varieties.)

A study by Voet (1989) on the adoption of innovations in the Gaza Strip, suggests that under conditions of financial support (subsidized loans, or grants) and proper agricultural extension, the adoption of modern irrigation technologies is a smooth process. According to Awartani (1990) in 1989/90 42 percent of fruit orchards and 97 percent of vegetables in Gaza Strip were irrigated by modern technologies. The corresponding figures for the West Bank are 16 and 33 percent for fruit orchards and vegetables respectively.

It should be noted that Awartani in his projections for agricultural demand for water in the West Bank and Gaza Strip already took into account modernization of the irrigation systems. More information is needed to evaluate the potential for further water saving by modern irrigation technologies in these regions.

A questionnaire on the potential for water saving in Israel with the aid of modern equipment suggests that it is quite limited (Spector and Yaron, 1993), because in most regions and for most crops, such equipment has already been installed. In the seventies the government of Israel ran a campaign aimed at saving irrigation water through modernization of the equipment. The campaign was supported by grants and subsidized loans and gradually achieved its declared goals. The questionnaire results also suggest that there exists a non-negligible potential for improving the management and the efficiency of irrigation by better scheduling, better timing and more accurate quantities of water applications.

"Greenhouse technology" is another potential avenue for water saving via substitution of capital and know-how for water. Comprehensive comparisons are difficult to make because there is a whole spectrum of greenhouse types as well as open field technologies with plastic covering of crops. Furthermore, greenhouse products differ significantly from open field technology products in terms of quality and season of supply. With these reservations in mind, some figures are presented in Table 14.5 to provide an order of magnitude of the difference in water use by the two technologies.

Finally, there is potential for water saving in agriculture through raising water prices for farmers. In Israel water for irrigation is allocated according to institutionally determined quotas; about 65 percent of the water is supplied by Mekorot, Israel Water Co. The prices are determined by the government, according to block differential policy, namely a lower price for the first part of the quota and a higher one for the rest. A high "penalty" level price exists for water use, which exceeds the quota. It is possible to introduce additional water saving incentives to farmers by modification of the current pricing system, i.e. raising the prices of water to reflect better their scarcity value. Various modification alternatives are being discussed; they differ in the relative weight assigned to economic efficiency versus income distribution considerations and the means for compensating farmers. The details of this discussion fall beyond the scope of this chapter. The interested reader is referred to Hochman and Hochman (1991), Kislev (1992), Yaron (1991) and Zusman (1991).

Table 14.5. Quantity of water per ton of product (m^3/ton)

	Greenhouse[a]	Open field[b]
Tomato	50	50-90
Cucumber	100	125
Melon	57	83-125
Eggplant	57	87-150

[a] Based on Yolk and Yaron (1992)
[b] Based on Viezel and Tsur (1990)

4. NEW WATER SOURCES FOR THE REGION

It is evident that in the future (the exact time depends on the population growth in Israel, the West Bank and Gaza Strip) the whole potential supply of fresh (high quality) water will be consumed by domestic use, with perhaps some minor allocation to high value crops sensitive to salinity[1] and other parameters of water quality.

In view of such a projected situation (the scarcity of water in Jordan and North Sinai, and the prospects for peace in the region) new potential sources of water supply for the region are being discussed. The underlying rationale is threefold: (a) solving the water problems in the region, (b) creating channels of regional cooperation, and (c) prospects for financial assistance from the international community in the spirit of promotion of peace in the region. Several large-scale projects are being discussed. They are based on either of the following: (1) desalination of Mediterranean sea water at sites close to the sea shore (e.g. at the junction of borders between Egypt, Gaza Strip and Israel, (2) conveying the Mediterranean Sea water or the Red Sea water to the Dead Sea (the Dead Sea is about 400 meters below sea level), generating energy and using it for desalination; (3) importing water to the region (e.g. Turkey). For details the interested reader is referred to Soffer (1992), Kally (1988), and Fishelson (1989). This paper does not attempt to evaluate the above alternatives for new source development.

5. MECHANISM OF WATER ALLOCATION TO ISRAEL AND THE PALESTINIAN ENTITY

In most cases water allocation among countries, regions within countries, and among alternative uses within countries, is subject to institutional-political decisions. Economists often point out the economic inefficiency of such institutional allocations. Policy makers who understand the claims of economists but do not accept economic efficiency as the dominating criterion also acknowledge it.

Orthodox economists are in favor of allocation via market mechanism (see e.g. Kislev, 1992; Zeitouni et al., 1994). This author counts numerous disadvantages in pure market mechanisms for water allocation within a country. Such a strategy may lead to (a) drastic changes and deviations from the status quo in agriculture, even though agriculture needs long-run supply reliability; (b) a conflict with national goals; (c) large scale projects that imply government intervention and coordination, especially in developing regions.

This author is in favor of an allocation-pricing-policy based on a mix of quota system with market mechanisms, with the latter applied only at the marginal segment of the quotas. Specifically the policy involves: (a) quota allocation system, adjusted gradually over the years; (b) block pricing differentials with the prices increasing for the higher shares of the quota; and (c) a substantially high "fine level" price for water used above the quota level. High water prices at the quota margins and gradual adjustment of the quotas over the years will increase the efficiency of water use over time and at the same time avoid drastic changes in allocation. It should be noted that differential block pricing of water is being applied in Israel.

Allocation of water between Israel and the Palestinian entity via competition and price mechanism might lead to even more severe difficulties than the effects of free competition for water within one country within 25-40 years. Depending on the growth of the population and the standard of living in the region, nearly all fresh high quality water from natural sources will be consumed for domestic use. Unless the income per capita in Israel and the Palestinian entity are similar, the price mechanism, under conditions of free competition, will lead to quite inequitable allocation of water for domestic uses between Israel and Palestinians. Similarly, either the Israeli or the Palestinian agriculture, or both, depending on the region and the product, might be subject to drastic shocks. Accordingly, in this case too, a system based on a mix of institutional-political allocation, with market mechanisms at the margin only, is suggested.

6. TOWARDS COOPERATIVE SOLUTIONS

This chapter is restricted to Israel, the West Bank and Gaza Strip, but the approach can be easily extended to include Jordan, North Sinai and other countries in the region.

A major conflict between Israel and the West Bank is over sharing the Mountain Aquifer water rights. The main Palestinian claim is that a major part of the Mountain Aquifer lies underneath the West Bank territory. However, regardless of the Palestinian viewpoint, from a hydrological and engineering vantage point, the appropriate place to pump this water is from the western part of the Coastal Plain, which lies within the boundaries of Israel. Historically, Israel has been using and pumping from this aquifer for wuite some time, a water volume roughly equal to its safe yield and therefore, claims riparian rights. It is not the purpose of this paper to review the international law on the issue of water rights from a transboundary groundwater; the interested reader is referred to Berberis (1991), Hayton and Utton (1989) and Benvenisti (1992). The bottom line of the analyses and interpretations of the law is that any settlement of this conflict should involve negotiations leading either to a solution or a scheme for a mechanism for reaching an agreement, such as arbitration, mediation, etc. Any agreement will take into account the abovementioned claims of the two parties, as well as their economic and social needs and their alternatives for the development of new water sources.

The situation with respect to Israel and the Gaza Strip is different. The southwest part of Israel (i.e. South Western Negev) and the Gaza Strip lie above the same coastal sub aquifer. The water requirements of both parties exceed the natural replenishment of this subaquifer. The supply of water to the South Western Negev is supplemented by importation of water from the northern parts of Israel; the water needs of the Gaza Strip could be met only by a similar importation of water from outside or from local desalination plants. The transboundary water rights conflict hardly seems to exist in this case; the case of the Palestinians' demand for additional water supply seems to be logically based on their genuine needs.

We apply concepts from cooperative game theory for designing a model, which might be helpful in the derivation of a cooperative solution. Two variants of the model are conceivable; they differ in the approach regarding the starting point for the analysis, (i.e. the non-cooperative solution) in terms of water allocation among the parties.

Variant A assumes that the starting point is represented by the status quo in the use of water. This, however, may be too far away from a realistic situation.

Variant B assumes that the starting point for the analysis is represented by a politically derived settlement, and the model is used for improvements and in refinements, both at the stage of finalizing the agreement and for modification over time.[2] In the following, without loss of generality, we refer to Variant B.

7. THE NON-COOPERATIVE SOLUTION

The analysis is static and refers to the first year of the agreement. From this year on it could be updated as the need arises depending on the institutional arrangements between the parties.

The starting point of the analysis for a given year is based on the following assumptions: 1) Water (pumping) rights of each party are predetermined by the political agreement. Urban consumption gets priority, the agricultural needs are not fully satisfied; 2) Sewage treatment and disposal arrangements are predetermined; 3) Each party (i.e. Israel, West Bank and Gaza Strip) develops its own additional water sources; 4) Each party aims to maximize its income derived from water use in agriculture subject to water availability, the availability of other resources and the prevailing technology.

A schematic presentation of the model for one party follows.

(1) Maximize Y_p $\qquad p = 1, 2$

Subject to income definition function

$$(2) \qquad Y_p = f\left(W^q_{pr},\ PC\left(W^q_{pr}\right),\ CC\left(W^q_{pr,\ r^*}\right) \ldots \middle|\ K_p\right) \qquad r \in R_p$$

and subject to other restrictions as listed:

(3) natural resource availability (*p*, *r*), technology (water supply, agriculture (*p*, *r*), urban demand for water, domestic and industrial (*p*, *r*),[3] water conveyance restrictions (technical and economic) (*p*, *r*), environmental restrictions (*p*, *r*), lower bounds on regional water supply for agricultural uses (*p*, *r*),[4] where

p = index of the party (Israel and the Palestinian entity), $p = 1, 2$;
r = index of the region within the *p*th party, with R_1 and R_2 being the sets of regions for the two entities, respectively;
q = index of water quality, $q = 1, 2, 3$, fresh, treated waste and brackish water, respectively,
Y_p = income generated by water ($p = 1, 2$);
W^q_{pr} = volume of water of quality q allocated to region r, party p;
PC = pumping cost per one water unit;
C = conveyance cost of one unit of water from region r to region r^*.

Distinction is made between three qualities of water ($q = 1, 2, 3$), fresh (high quality) water, treated wastewater and brackish water, respectively. In a more detailed analysis further refinements of the definition of treated wastewater and brackish water could be introduced according to the level of treatment of wastewater and the salt content of brackish water.

The optimal solutions of this model, separately for $p = 1$ and 2 show the allocation of water within each party, the agricultural production and income by regions given the “non-cooperative” arrangements. The degree of detail depends on the choice of the model builder. Such a model has been recently constructed for Israel by the author and collaborators. It contains a water supply and demand sections (mainly agricultural). A separate section refers to 14 agroclimatic regions and 25 irrigated products in each region (fruits, vegetables and field crops) and their demand for local consumption, export and processing. This is a large interregional competition linear programming model. The technical problems are minimal; the major effort is the preparation of data.

8. DERIVATION OF THE COOPERATIVE SOLUTION

The following assumptions underlie the cooperative solution: (1) joint use of natural water sources; (2) joint development of new water sources; (3) joint planning and control of sewage treatment, disposal and/ or use of wastewater.

A formal presentation of the cooperation model follows. It aims at the derivation of the efficiency frontier in the two parties' income space with water allocation being the major decision variable.

(4) Maximize $\overline{Y}_1^{coop}$

Subject to

(5) $\overline{Y}_2^{coop} \geq YK_2$ with YK varying parametrically from zero to a maximal feasible level

(6) $$\overline{Y}_p^{coop} = \begin{matrix} g\left[W_{pr}^q,\ PC\left(W_{pr}^q\right),\ CC^{**}\left(W_{pr,\ r^{**}}^q\right) \ldots \middle|\ K_p\right] \\ r \in R_p,\ r^{**} \in R_1 \cup R_2 \end{matrix}$$

and

(7) Restrictions set like (3) modified to suit cooperation (e.g. inter-party conveyance, inter-party environmental issues, etc. Note that the allocation of water for domestic uses by the two parties is predetermined.)

where:

$\overline{Y}_p^{coop}$ = the income of party p under cooperation, and

CC^{**} = the conveying cost of one unit of water from region r of party p to any other region r^{**} of party 1 or 2 ($r \in R_1UR_2$); and other symbols being as above.

By parametrically varying YK_2 the efficiency frontier in the Y_p income space (= income transformation curve) can be derived.[5]

The income efficiency frontier is presented in Figure 14.1, with the axes being ΔY_1 and ΔY_2, the additional income derived over and above that derived with no cooperation. Each point on the efficiency frontier represents a given allocation of water among the regions of the two parties, the income, and the major crops (to any desired/degree of detail).

Point *Q* in Figure 14.1 represents the "non-cooperative" solution achieved via the political agreement, and the related derived incomes. The efficiency frontier curve *AB* represents the improved income combinations due to water reallocations and joint development of new water sources.

Which point on the efficiency frontier curve is the "best" and who makes the choice? Before dealing with this issue, policy alternatives regarding the distribution of income derived from water between the two parties should be considered.

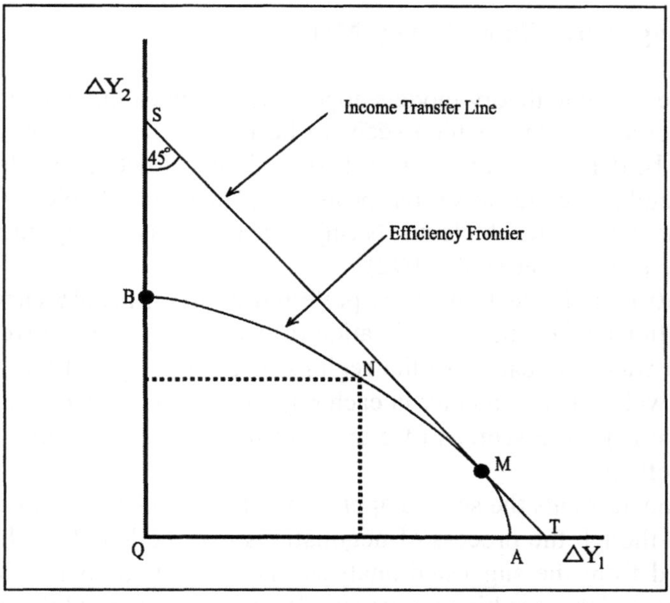

Figure 14.1. Income efficiency frontier: Israel and the Palestinian Entity

8.1. Water Allocation and Income Distribution Policies with or without Side Payments

Two major policies regarding distribution of incomes are open to the cooperating parties: (1) The cooperation is restricted to the development of new sources and the exchange of water quotas determined by the *a priori* political agreement. (2) The cooperation involves new water source development, water quota exchange and side payments, i.e. direct income transfers.

If the side payments policy is agreed upon, the two cooperating parties, will maximize their joint income at point *M* in Figure 14.1 and then redistribute the income along the income transfer line *ST*. The implication of direct income transfers is that income generated by one party is transferred to the other one, without direct reference to the quantity and the price of water. Such a policy, while increasing the welfare (*ST* lies above *AB)* could raise criticism on the grounds of disproportion between the quantities of water and the amount of money to be transferred, and be politically unacceptable.

The less efficient, and arguably the more politically acceptable solution, is to restrict the set of inter-party reallocations, under cooperation, to the efficiency frontier curve *AB*.

Which point on the efficiency frontier curve *AB* should be chosen? Two possible approaches to this issue are discussed below.

8.2. Cooperative Game Theory Models

The use of game theory models appears to be appealing. Once the parties agree on certain axioms, which seem to be reasonable, the solution can be technically be derived on the basis of these axioms. However, models used by the author and associates in similar problems (The Core, Shapley Value, Generalized Shapley Value, Nucleous, Nash) were generally disappointing (Yaron and Ratner, 1990; Dinar et al., 1992).

Straightforward selection of the preferred point on the *AB* curve, following negotiations, takes into consideration (a) water allocation between the two parties; (b) water allocation to the regions of each party; (c) marginal value product of water of each quality in each region; (d) water-generated income in each region; (e) the essentials of crops mix in each region, and (f) the environmental effects.

By some accounts the second approach is likely to lead to sensible allocations, even though the process of negotiations may be lengthy. The information derived from the suggested analysis may provide a sound basis for the negotiations and the resulting agreement. It should be recalled that the starting point for the analysis—the non-cooperative solution—is also derived (according to our assumptions), through a negotiation process, so that the mechanism for negotiations already exists. As previously mentioned, it is also possible to apply the suggested approach beginning with the status quo ante according to Variant A. In such a case it could be used interactively with the negotiations at the two stages regarding (a) the agreement on the non-cooperative solution, and (b) the choice of the preferred point on the income efficiency frontier. Note that the choice of the "preferred point" also provides the allocations of water between the two parties and their respective regions. These can later be subdivided by each party to the generic units in each region such as water cooperatives, farms, etc.

8.3. Incorporation of Price Mechanism at the Margin of the Institutional-political Water Allocations

By definition, the income transformation curve *AB* is the locus of Pareto optimal points, namely the income of one party cannot be increased without decreasing that of the other one. However, the curve *AB* is derived using an aggregate model and approximate average or "typical" input-output data. Furthermore, the dimensions of such a model imply that only the major crops in each crop category can be included and a group of "miscellaneous" must be defined for each category of crops (fruits, vegetables, field crops). Therefore the model is useful for deriving approximately optimal allocations of water among the two parties and their respective regions, but there is necessarily, some discrepancy between the model data, the results and reality.

The model results can be improved in the real world by the introduction of interparty, interregional and interfarm water mobility affected by a price mechanism with respect to a restricted segment of water allocations.

The actual way of incorporating a price mechanism as a driving force, activating water mobility, is a major problem in itself. It could be incorporated via block differential water prices as currently practiced in Israel or via some kind of auction. The problem is complicated mainly due to (a) restricted physical capacity for water conveyance, and (b) the difference in the level of development of the Israeli and the Palestinian economies at least in the near future. At this point of time the author refrains from suggesting the details of such a mechanism.

The suggested approach originates in the conviction of the author that: (a) any decision regarding water allocations must take into consideration the details of its implications with respect to income, MVP of water, national goals, environmental effects etc.; (b) the political-institutional allocations are economically inefficient; (c) allowing only the market mechanism to determine the allocations might lead to drastic changes and politically unacceptable solutions; (d) a mix of political-institutional allocations with a market mechanism at the margin of water allocation is needed.

REFERENCES

Awartani, H. Projection for the demand for water in the West Bank and Gaza Strip. Unpublished paper prepared for the World Bank. Ben Gurion University. Mimeo. 1990.

Benvenisti, E. International law and the mountain aquifer (Paper presented at the First Israeli-Palestinian International Academic Conference on Water). Zurich, Switzerland. 1992.

Berberis, J. The development of international law of transboundary groundwater. *Natural Resource Journal,* 24:31-167. 1991.

Dinar, A., A. Ratner and D. Yaron. Evaluating cooperative game theory in water resources. *Theory and Decisions,* 31:1-20. 1992.

Fishelson, G. The economic benefits of peace in the Middle East, Research Report. The Armand Hammer Fund for Economic Cooperation in the Middle East. Tel-Aviv University, Israel. Mimeo. 1989. (Hebrew)

Goldberger, S. The potential of brackish water in Israel. Document presented to the Committee on Brackish Water. Israel Water Commission. Israel. 1992.

Hayton, R. D. and A. E. Utton. Transboundary groundwaters: The Bellagio Draft Treaty. *Natural Resource Journal,* 26:663-722. 1989.

Hochman E. and O. Hochman. A policy for efficient water pricing in Israel. *Economic Quarterly,* 150:502-523. 1991. (Hebrew)

Kally, E. Biterritorial sea water project, Feasibility Study Report. Ramat Hasharon, Israel. Mimeo. 1988. (Hebrew)

Kally, E. "Water in Peace." Sifriat Poalim Publishers and Tel Aviv University. Tel Aviv, Israel. 1989. (Hebrew)

Kislev, Y. The water sector - basic concepts and applications Center for Agricultural Economic Research. Rehovot, Israel. 1992. (Hebrew)

Moore, J. M. An Israeli-Palestinian water sharing regime (Paper presented at The First Israeli-Palestinian International Academic Conference on Water). Zurich, Switzerland. 1992. Truman Institute, The Hebrew University.

Nvo, N. Comments Paper presented at the 7th Continuing Workshop on Israel Water Issues. Center for Agricultural Economic Research. Rehovot, Israel. 1992. (Hebrew)

Regev, A., Jaber, R. Spector and D. Yaron. Economic evaluation of the transition from a traditional to a modernized irrigation project. *Agricultural Water Management,* 18:347-363. 1990.

Schwartz, Y. Israel water sector study. Ben Gurion University, Mimeo. 1990.

Shuval, H. I. Approaches to finding an equitable solution to the water resources problems shared by Israel and the Palestinians over the use of Mountain Aquifer in: "Water Conflict or Cooperation?" (G. Baskin, ed.). Israel Palestinian Center for Research and Information. Jerusalem, Israel. 26-53. 1992.

Soffer, A. "Rivers of Fire: The Conflict of Water in the Middle East." Am-Oved Publishers Tel Aviv, Israel. 1992. (Hebrew)

Spector, R. and D. Yaron. Summary of Questionnaire on irrigation systems in Israel (Unpublished). Department of Agricultural Economics. The Hebrew University of Jerusalem. Rehovot, Israel. 1993.

Tahal, Consulting Company. Water Master Plan (Interim report). Tel Aviv, Israel. 1988. (Hebrew)

Viezel, L. and Y. Tsur. Computerized cost estimates of vegetable growing. Ministry of Agriculture Extension Service. Tel Aviv, Israel. Mimeo. 1990. (Hebrew)

Voet, H. Adoption of innovations on private Arab farms. Paper presented at the Trinational Egypt-Israel-USA Project Conference. Alexandria, Egypt. December, 1989.

Volk, A. and D. Yaron. Review of technologies, cultivated areas and economic indicators in vegetables (Research report). The Hebrew University of Jerusalem. Rehovot, Israel. 1992.

Yaron, D. Allocation of water and water prices in Israel. *Economic Quarterly,* 150:465-478. 1991. (Hebrew)

Yaron, D. and A. Ratner. Regional cooperation in the use of irrigation water: efficiency and income distribution. *Agricultural Economics,* 4:45-58. 1990.

Yaron, D. and A. Regev. Is modernization of traditional irrigation systems in arid zones economically justified? In "Proceedings International Conference on Irrigation: Theory and Practice." Southampton University. United Kingdom. 1989.

Zeitouni, N., N. Becker and M. Shechter. Models of water market mechanisms and an illustrative application to the Middle East. *Resource and Energy Economics,* 16:303-319. 1994.

Zusman, P. A conceptual framework for a water resource regulation policy in Israel. *Economic Quarterly,* 150:440-464. 1991. (Hebrew)

---

[1] The bulk of the supply to agriculture would be treated wastewater (it is assumed that in Israel about 60 percent of water domestically consumed can be treated and used in irrigation).

[2] Examples of ideas for possible political settlements can be found, e.g. in Shuval (1992) or Moore (1992).

[3] It is assumed that urban demand gets priority in water supply and it is predetermined.

[4] The aim of these restrictions is to prevent drastic changes and situations of "drying out" agricultural regions.

[5] Actually with YK_2 varying parametrically over discrete values, the efficiency frontier is an upper boundary of a polyhedron, with the vertices each representing a particular solution. The segments connecting the vertices are convex combinations of the solutions.

15

PLACING DAN YARON'S WORK IN THE LITERATURE

David Zilberman
University of California, Berkeley, USA

Ariel Dinar
World Bank, Washington DC, USA

Dan Yaron's long and productive academic career created a significant body of work that shaped and influenced both the research agenda and the evolution of the analytical frameworks of agricultural, natural resources, and environmental economics. Yaron's work integrated creativity with pragmatism, rigor, and relevance. Though most of Yaron's work was inspired by problems he encountered while working and living in Israel, he had a unique ability to generalize these problems in lessons that became useful for people from all over the world. He became a world-leading expert on water resource management issues and his agenda and methods advanced new scientific knowledge. As the arsenal of economic tools grew, he took advantage of new techniques and adapted them to provide insights on some of the most important and difficult resource management problems of our time—limited quantity and poor quality of water.

The area of environmental and resource economics has become a vibrant, exciting, and important area of economic research. This relatively young field gained its distinct identity in the early 1970s. The energy crisis and the establishment of the Environmental Protection Agency (EPA) are two major events that link the emergence of environmental and resource economics as a distinct subfield. The intellectual foundation and research methodologies of environmental economics, however, were previously established.

Environmental and resource economics owes much to visionary and intellectual giants such as Hotelling, and they have continued to expand methodologies and knowledge accumulated by researchers in the areas of fisheries and forest economics.

Agricultural economics has contributed to the massive base of theoretical knowledge, empirical methodology, and institutional foundation applied to the environmental and resource economics subfields. Ciriacy-Wantrup and Castle have emerged as leaders within these subfields and have contributed greatly. The work of agricultural economists, however, has been most applicable to environmental and resource economics in the area of water. Agricultural economics has co-evolved with environmental and resource economics since the establishment of the latter. While each subfeild has a unique institutional peculiarity and thematic emphasis, there is a tremendous amount of cross-fertilization in the field of water.

Dan Yaron and his contemporaries—Burt, Cummings, Young, Martin, and Gardner—are among the leading agricultural economists who laid the foundation of modern water economics research. Burt (1964) as well as Cummings and McFarland (1974) anchored water economics within the general theory of resource management and introduced novel approaches to analyze complex problems such as conjunctive uses of ground and surface water and seawater intrusion into aquifers. Young and Martin (1969) used economic analysis to challenge the water development paradigms of the 1950s and 1960s and economic reasoning to halt the expansion of new water projects. Gardner (1983) initiated much of the research on the impact of water institutions on water resource management and provided critical intellectual ammunition for the introduction of markets to allocate water.

Dan Yaron's contribution was in the area of water resource management. He was unique in his capacity to develop quantitative methods to optimally determine how to manage water at several levels, starting with plants, continuing to fields in farms, and finally reaching regional and national levels. He was also responsible for shifting the emphasis to quantitative analysis of water quality problems and addressing issues of externality associated with water resource use. More importantly, he established and embodied a true interdisciplinary research program, collaborating with soil scientists, meteorologists, crop scientists, and others.

Many environmental and ecological economists aspire to do multidisciplinary research, but few succeed. Yaron showed, for example, that multidisciplinary research entails, not only cooperating with members of other disciplines to develop integrated products, but also incorporating the basic knowledge of biology and physical sciences into an economic decision-making framework. Yaron used relationships obtained by natural scientists in modeling production and growth processes and this knowledge provided realism and vitality to his quantitative analysis. Some of the chapters in this book demonstrate that the economics and soil sciences are fused into one integrated, decision-making framework. Using his knowledge and skills, Yaron transformed the boundaries between economics and other agricultural sciences and, thus, warrants being referred to as both an integrated water and environmental scientist.

One of Yaron's amazing achievements as an educator was that he imparted his multi-disciplinary approach to his students in Israel. Yaron and his students formed a center of excellence with a holistic approach to water economics and management. A similar group emerged in the 1970s in Riverside, California, under the leadership of Vaux, Jr. and Letey. The interaction between these two groups has been very important in developing a critical mass of research on the economics and management of water and water quality, which embodies the interdisciplinary spirit that Yaron pioneered. This holistic approach has been pursued productively in many research centers in the United States, Australia, and elsewhere. Research on water has become an area of environmental and agricultural economics where economists' agendas and research efforts are most integrated with other disciplines.

Dan Yaron received his PhD from Iowa State University. He was one of Earl O. Heady's outstanding students, and his approach to research has been heavily influenced by Heady's work and methods. Heady's major achievement was the up-scale application of modern techniques, especially linear programming. He took advantage of the immense capacity of computers to solve agricultural production and management problems. Yaron relied on, and significantly improved, the approach introduced by Heady and Egbert (1959), where statistical tools were used to estimate some of the basic relationships of agricultural production systems. These parameters were used in programming models to determine and analyze resource allocation.

Yaron brilliantly applied Heady's original approach to water allocation problems (see Chapter 4). Due to the importance of citrus and perennial crops in Israel, Yaron emphasized the use of dynamic models and techniques for applied research, and he also recognized the importance of uncertainty considerations in resource allocation. Thus, he creatively adjusted his models and analyses to take into account both random events and dynamic factors that affect agriculture. Several chapters in this book (in particular, Chapters 6, 7, and 8) demonstrate Yaron's ingenuity in applying advanced tools of operational research (Markov chain analysis, stochastic dynamic programming, etc.) to solve water resources and environmental management problems.

The mathematical programming techniques that Heady popularized, and which Yaron and others expanded and applied to soil and agricultural resource management problems, are normative in nature. They aim to find optimal resource allocation subject to constraints. However, many of the applications of this model have been for predictive purposes. Takayama and Judge (1971), for example, exposed and expanded the capacity of these models to design and analyze resource allocation over space. Heady and his colleagues in the Center of Agricultural and Resource Development, at Iowa State University developed a unique capacity to assist government agencies in the design of policy and impact assessment.

The use of programming models for predictive purposes has been the subject of major controversy. According to theory, economic agents are optimizers and maximize profit or utility; thus, it seems that the use of optimization models to predict behavior is reasonable. However, as Friedman (1953) noted, models are only an approximation of reality. Econometric techniques were developed for statistical use to systematically capture the delineation between actual behavior and terms that are expected by theory. Calibrating mechanisms have been used to adjust results of programming models to actual reality, and purists held some of the ad hoc features of calibration suspect. Furthermore, the fixed coefficient production functions that were assumed in many of the programming models were another objectionable property of many of the programming exercises that Yaron and his colleagues were engaged in.

In spite of an occasional criticism, Yaron and others have persisted with the development and application of programming models. In retrospect, their tools of analysis have been immensely successful. They were able to capture the heterogeneity of economic systems that had not been incorporated. They provided an excellent vehicle for interdisciplinary dialogues. Most importantly, they passed the test of "revealed preference." Government agencies and decision-makers use programming models to make their own predictions. Over the years, their performance and methodological underpinnings have been constantly improved, and programming models have played a major role in analyzing important environmental and agricultural policy problems. Howitt (1995), in his water allocation analysis, and McCarl et al. (2000) in their study on the impact of climate change, have followed and improved the pioneering effort that Heady started and Yaron and others expanded. Furthermore, using the entropy techniques, Paris and Howitt (1998) developed methodologies that provided statistical framework models. Moreover, econometric studies that were inspired by putty-clay models have shown that the assumption of fixed production coefficients that are so prevalent in programming models may be quite realistic in the short run (Just et al., 1990). In retrospect, the programming models that Yaron applied and championed improved and become the major workhorse of applied agricultural and environmental economic research, in both policy design and impact assessment.

Dan Yaron's research agenda was inspired by the water problems in Israel, and he raised the literature's attention to important problems that might have otherwise been underemphasized by economists. An example is the management of water quality, especially for salinity and irrigation at the regional level. The analytical framework for the demand of irrigation management with variations in water quality is presented in Chapter 7 in this book. Many of the results were presented years later in Rosen's (1974) seminal paper on hedonic pricing in a general format. Yaron's major conclusion was that all water should not be viewed the same but, rather, in terms of its many characteristics and be priced accordingly. He determined that there should be a premium on water based on its unique quality, and consistent with hedonic pricing (which was later formally introduced to economic literature). Yaron derived hedonic, implicit prices of quality for water in his work. He introduced trade-off relationships between quality and quantity and consistent with hedonic pricing of irrigation water and obtained optimal irrigation schemes that specified how much water of various qualities to use under different conditions.

Yaron's work preceded a large literature on discriminatory treatment of various qualities of water. Some later studies, for example, Rhodes and Dinar (1991) expanded Yaron's original ideas and developed schemes where water quality in crop irrigation varies throughout the season. They also suggested separate inventories for water with different qualities. Studies that originated in California (Caswell et al., 1990) emphasized the importance of land quality in irrigation and drainage management. Dinar and Zilberman (1991) suggested that crop selection, technology choice, and water use were all functions of water and land qualities. Chapter 6, for example, illustrates Yaron's emphasis on the value of water, not only dependent on quality, but also on timing throughout the season. The work of Griffin and Hsu (1993) formally derived the assessment of water as functions of location and use. Thus, the analysis of water pricing has advanced from Yaron's early work, where he relates the price of water to its quality, to a more holistic analysis where water is assessed according to quality, the quality of land on which they are applied, and where, when, and how it is used.

The analysis of the impact of timing and quality on water use and value was part of a major theme of Yaron's work, namely, that resource management is done under conditions of heterogeneity. Assumptions of "representative agents" that typify much of the theories that provide guidelines for intervention to address issues of externalities and market failure do not hold in reality. As an empiricist, Yaron recognized the complexity of economic systems and the inefficiency that may result from introducing a uniform policy when heterogeneity exists. That is apparent in his work on the water economy of Israel and the inefficiency associated with using a uniform water price throughout the country (Chapters 2 and 4). Yaron emphasized the importance of introducing differentiated pricing and other policy tools to address varying conditions. However, in his discussion and analysis of results, he recognized the managerial challenge of implementing differentiated policy schemes. He suggested that the gains associated with adjustment to variability would be balanced with the extra implementation cost to determine optimal policy design.

A review of Yaron's body of work suggests that he constantly updated and diversified the set of tools he applied in analysis. His work moved from linear programming to dynamic and stochastic programming. It seems that Yaron's continuing "self-improvement" was associated with his mentoring efforts, since some of his more technically advanced work was done mostly with his students (which is the way it should be). He used cooperative game theory (see Chapter 12) to obtain decision rules to allocate resources among competing groups with different power. Over the years, his work emphasized more the stochastic aspect of decision making as is apparent in Chapter 9. He recognized that decision rules were made under conditions of uncertainty and suggested that water policy management be evaluated by its statistical reliability and power. Namely, policymakers should recognize and take into account the likelihood that policies often do not achieve their desired results. This line of research has advanced over the years as political economic models have been introduced to analyze water allocation disputes (Rausser and Zusman, 1991).

One interesting co-authored econometric piece, which makes a significant contribution to the literature, is on adoption of irrigation technology (Chapter 13). This is part of a rich body of literature (see surveys by Caswell, 1991 and Green et al., 1996). These studies identified factors (prices, land quality, water supply shortages, etc.), which led to the adoption of advanced irrigation technology. Yaron's work recognized that technologies are born and then die, and he took an interesting look at the dynamics of transition among technologies. Factors that lead to the introduction and extension of technologies also lead to its demise.

Dan Yaron developed a global and strategic perspective on water that was most apparent in his analysis of the Israeli water economy. He used his rich base of quantitative knowledge to develop a prescription that incorporated efficient pricing with the use of advanced technological fixes, enabling neighbors in the region to utilize resources to enhance economic growth and prosperity. Yaron warns against the waste and inefficiency associated with uniform prices that ignore differences between water cost and productivity and the consequences of continuous deterioration of water quality. He suggests that Israel use agricultural water for profitable crops and predicted a joint management of irrigation and reclaimed water.

Dan Yaron's work has had a lasting influence on the economics of water and natural resources in general and established an intellectual base of an interdisciplinary policy perspective on water resource management in Israel and around the world. His written work is part of a larger legacy that includes his students and the long-lasting impression he left with many of us.

REFERENCES

Burt, O. R. The economics of conjunctive use of ground and surface water. *Hilgardia*, 36:31-111. 1964.

Cummings, R. G. and J. W. McFarland. Groundwater management and salinity control. *Water Resources Research,* 10:909-915. 1974.

Caswell, M. F. Irrigation technology adoption decisions: empirical evidence. "The Economics and Management of Water and Drainage in Agriculture" (A. Dinar and D. Zilberman, ed.). Boston, MA: Kluwer Academic Press. 1991.

Caswell, M., F, E. Lichtenberg and D. Zilberman. The effects of pricing policies on water conservation and drainage. *American Journal of Agricultural Economics,* 72:883-890. November, 1990.

Dinar, A. and D. Zilberman, The economics of resource-conservation, pollution-reduction technology selection: the case of irrigation water. *Resources and Energy*, 13:323-348., 1991.

Friedman, M. "Essays in Positive Economics." Chicago, IL. University of Chicago Press. 1953.

Gardner, B. D. Water pricing and rent seeking in California agriculture. "Water Pricing and Rent Seeking in California Agriculture." Cambridge, MA. Ballinger Publishing Company. 1983.

Green, G., D. Sunding, D. Zilberman and D. Parker. Explaining irrigation technology choices: a microparameter approach. *American Journal of Agricultural Economics*, 78(4):1064-1072. November, 1996.

Griffin, R. C. and S. H. Hsu. The potential for water market efficiency when instream flows have value. *American Journal of Agricultural Economics*, 75:292-303. 1993.

Heady, E. O. and A. C. Egbert. Programming regional adjustment in grain production to eliminate surpluses. *Journal of Farm Economics*, 41:718-733. 1959.

Howitt, R. Positive mathematical programming. *American Journal of Agricultural Economics*, 77:329-342. May, 1995.

Just, R. E., D. Zilberman, E. Hochman and Z. B. Shira. Input allocation in multicrop systems. *American Journal of Agricultural Economics,* 72 (1):200-209. February, 1990.

McCarl, B. A., D. M. Burton. D. M. Adams, R. J. Alig and C. C. Chen. Effects of global climate change on the U.S. forest sector: response functions derived from a dynamic resource and market simulator. *Climate Research*, 15(3):195-205. 2000.

Paris, Q. and R. E. Howitt. An analysis of ill-posed production problems using maximum entropy. *American Journal of Agricultural Economics*, 80, 1:124-138. February, 1998.

Rausser, G. C. and P. Zusman. Organizational failure and the political economy of water resource management. "The Economics and Management of Water Drainage in Agriculture" (A. Dinar and D. Zilberman, eds.). Boston, MA. Kluwer Academic Press. 1991.

Rhodes, J. D. and A. Dinar. Reuse of agricultural drainage water to maximize the beneficial use of multiple water supplied for irrigation. "The Economics and Management of Water and Drainage in Agriculture" (A. Dinar and D. Zilberman, eds.). Boston, MA. Kluwer Academic Press. 1991.

Rosen, S. Hedonic prices and implicit markets: product differentiation in perfect competition. *Journal of Political Economy,* 82(1):34-35. January, 1974.

Takayama, A. and Judge, G. G. "Spatial and Temporal Price and Allocation Models." Amsterdam: North-Holland. 1971.

Young, R. A. and W. E. Martin. The need for additional water in the arid southwest: an economist's dissent. *The Annals of Regional Science*, 3(1). June, 1969.

APPENDIX: DAN YARON'S GRADUATE STUDENTS[1]

1960 Moshe Ben David
Conclusion of an investigation into milk-producing farms in Moshav Talmai Yechiel

1962 Dan Arnon
Economic evaluation of irrigation water quality-methodology and empirical evaluations

1963 Gideon Fishelson
Local analysis of the Israeli fruit sector –application of linear programming
Yakir Plessner
A national planning model by regions of field crops and vegetables
Naftali Prigort
Marginal productivity of water in the "Nir Shalhin" region of the Western Negev

1964 Avraham Sabotnik
Estimates of response functions to irrigation water in selected crops and their incorporation into the programming of a moshav farm in the Northern Negev

1965 Dan Marom
Application of linear programming and economic analysis of a kibbutz farm in the Gilboa Region
Avraham Oron (Simcha)
Marginal productivity and demand for water of moshav farms in the Taanach Region
Mordechai Weisbrod
The credit system and the rural farm-activity and duality under uncertainty
Moshe Weiss
Optimal production programs and estimates of marginal productivity of water in kibbutz farms in the Northern Negev

1966 Dan Frumkin
Optimal water allocation on a kibbutz farm in the Gilboa Region

1972 Amikam Olean
Economic evaluation of saline water in irrigation-methodology and empirical application

1974 Avraham Polovin

Farm plan and how it changes after use of saline irrigation water

Igal Samid

A simulation model of wheat irrigation under uncertain rainfall as part of a comprehensive planning of field crops in a kibbutz farm.

Gad Stratiner

A simulation model for economic evaluation of wheat under uncertain rainfall

1975 Avraham Miloh

Economic analysis of efficient use of saline water in the Bait Sheaan Valley.

1978 Rami Raviv

Estimating the damage to a kibbutz farm in the Western Jeesrael Valley due to possible increase in irrigation water salinity

1979 Benyamin Harpinist

A model for optimal timing and allocation of saline irrigation water

1980 Eli Feinerman

Economics of irrigation with saline water under conditions of uncertainty (PhD dissertation)

Dov Golan

Analysis of factors for success and failure of industrial enterprises in kibbutz farms

1981 Arie Sadeh

Models for optimal allocation of water during pick season

1983 Aaron Ratner

Economic evaluation of regional cooperation in water use for irrigation-optimal allocation of water quantity and quality and selected income distribution

Anat Segev

Industrialization in moshav farms

1984 Ariel Dinar

Economic evaluation of the use of treated wastewater in agriculture and cost benefit allocation on a regional basis (PhD Thesis)

Doron Shavit

Analysis of the development trends of a moshav farm in the Sharon, using a financial-economic planning model

1985 Yochanan Weiler
Incorporating industry in a moshav-impact on finance and production

1986 Arie Regev (Blei)
Cost benefit analysis of modernization of regional irrigation project
Giora Gavri
Characterization of high-tech enterprises in kibbutz industries
Rivka Spector
Analysis of a vegetable family farm in the Central Arava Region under uncertainly

1988 Arie Volk
Factors affecting the growth of family debt in the moshav

1990 Efrat Hadas
Economic aspects of salinity and drainage in a kibbutz farm in the Jeesrael Valley
Avishai Rotenberg
Selection of optimal policy for selection of milking cows
Erez Wolf
Credit arrangements and joint liability in moshav farms

Year Unknown

Moshe Golan
Multi stage planning of wheat irrigation under uncertainty
Shlomit Hertz
Farming as part time-impact of risk and risk aversion
Binyamin Hofshi
Five year linear programming of kibbutz farm
Anat Keren
Characterization of high-tech industry and adaptation to conditions of moshav farms
Hana Savir
Financing, credit and guarantees for family factories in the moshav
Michael Friedman
Estimates of a production function and marginal productivity of irrigation water in moshav farms in the Negev
Shimon Shamla
Economic optimization of citrus irrigation with saline water and the value of additional information
Zeev Shuval
Integrated analysis of the water supply system and irrigation in a farm

Yano Tiroler

Decision tree model for kibbutz farm planning in the long-run under uncertainty

---

[1] Students' names and dissertation titles were translated from Hebrew by the editors. It is possible that several names have skipped our attention, and are not on this list.

INDEX

MIX
Papier aus verantwortungsvollen Quellen
Paper from responsible sources
FSC® C105338

If you have any concerns about our products, you can contact us on
ProductSafety@springernature.com

In case Publisher is established outside the EU, the EU authorized representative is:
Springer Nature Customer Service Center GmbH
Europaplatz 3, 69115 Heidelberg, Germany

Printed by Libri Plureos GmbH
in Hamburg, Germany